The Book of Daun Burnel the Ass

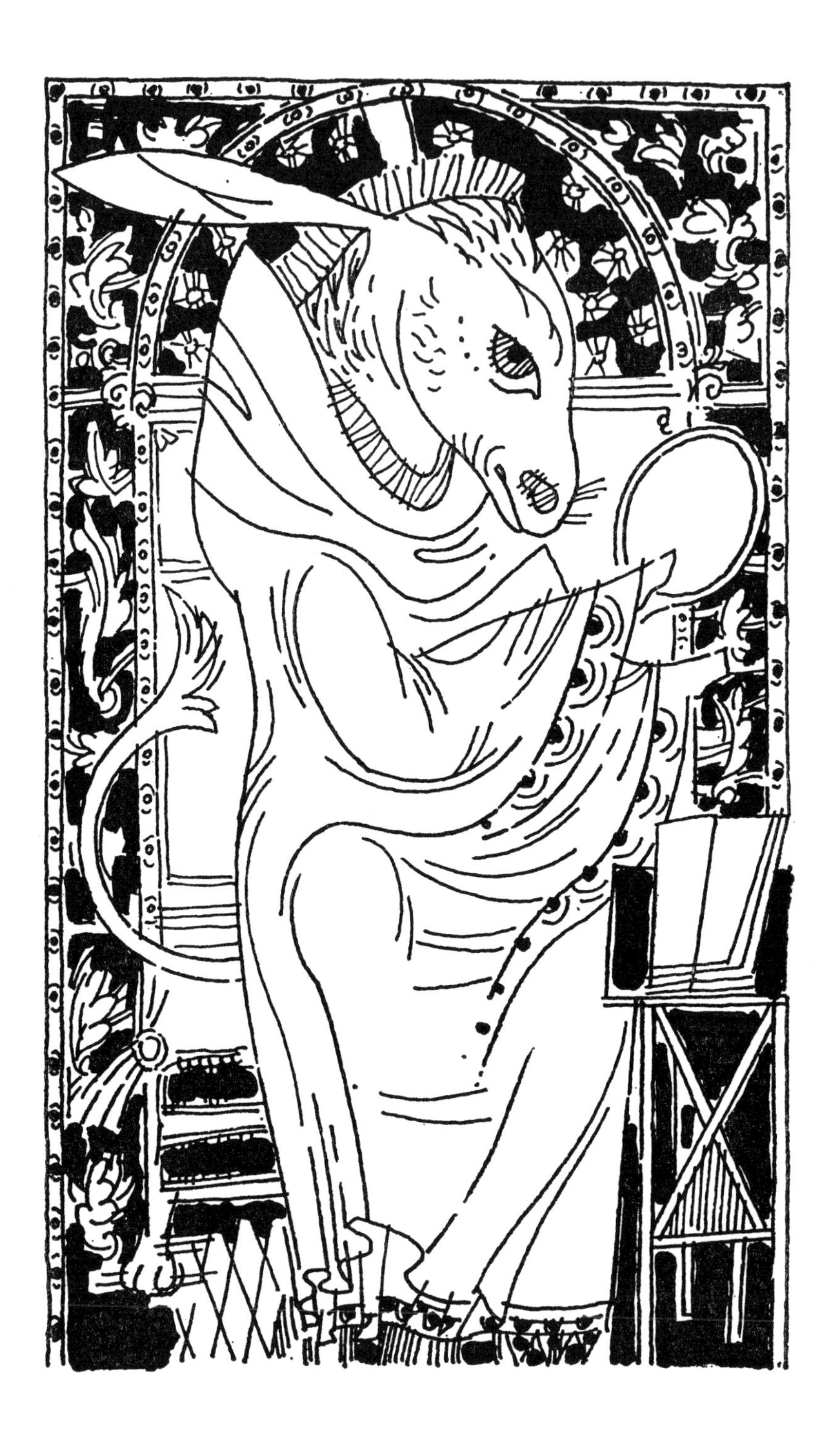

THE BOOK OF DAUN BURNEL THE ASS

Nigellus Wireker's *Speculum stultorum*

Translated with an Introduction and Notes by

GRAYDON W. REGENOS

Illustrated by LUIS EADES

AUSTIN · UNIVERSITY OF TEXAS PRESS

Library of Congress Card Catalog No. 59–8121

Published with the Assistance of a Grant
from the Ford Foundation
under Its Program for the Support of Publications
in the Humanities and Social Sciences

Manufactured in the United States of America
by the Printing Division of the University of Texas

To
Kenneth and Robert

Preface

The *Book of Daun Burnel the Ass,* as Chaucer calls Nigellus Wireker's *Speculum stultorum (The Mirror of Fools),* has never before appeared in an English translation. The reason why this lively and delightful twelfth-century poem has not received greater recognition in modern times must certainly be a lack of scholars in the field rather than any conception that the work does not have sufficient merit. Latin scholars have usually, and quite properly, concentrated their efforts on the great masterpieces of the classical period, but, in my opinion, they have generally paid too little attention to the vast amount of Latin written during the Middle Ages, the Renaissance, and later. During the past generation, however, it has been encouraging to note increasing interest and activity in the field of Late Latin, and although much has been done, there remains much more to be done. There is, for example, a need for more translations, not only to facilitate the work of the scholar in closely related fields, but to make these works available to the general reader. The present translation is offered as a small contribution toward the fulfillment of that need.

The idea of first attempting this translation was conceived a few years ago from a brief note in the *Classical Bulletin* by my fellow classicist, Professor Leo Kaiser of Loyola University of Chicago. His eloquent plea for translations of important Medieval and Renaissance authors such as Nigellus Wireker led me to the *Speculum stultorum.* I had long been familiar with this interesting work, but the thought of translating it had not occurred to me before. The task itself has given me much enjoyment.

I wish to acknowledge my deep debt of gratitude to my former teacher and friend, the late Professor Charles H. Beeson of the University of Chicago, who introduced me to the field of Medieval Latin literature and fostered my interest in the subject.

G. W. R.

Contents

The Book of Daun Burnel the Ass

INTRODUCTION

The Life of Nigellus Wireker

THE AUTHOR of the *Speculum stultorum,* or *Mirror of Fools,* is commonly known as Nigellus Wireker. The known facts of his life are rather meager, and even his surname is uncertain. It is known, however, that his period of activity fell within the last half of the twelfth century, during the reigns of Henry II and Richard I in England. Since he speaks of himself in the opening verse of his prologue to the *Speculum stultorum* as being an old man,[1] and there is good evidence that his work was probably written shortly before 1180,[2] some idea may be gathered as to the approximate date of his birth. It has been suggested that he may have been born about 1130.[3] He became a monk in Christ Church at Canterbury, perhaps before the murder of Thomas Becket in 1170, for he speaks of having had frequent contacts with the great archbishop and chancellor.[4] He is said to have been precentor of Canterbury, but I find no suggestion of this anywhere in his own writings.[5]

In his second major work, the *Contra curiales et officiales clericos,* which was apparently written in the early years of the last decade

1 "Suscipe pauca tibi veteris, Willelme, Nigelli."

2 It is assumed that the poem could hardly have been written later than 1180 because Louis VII of France, who died in that year, is mentioned.

3 M. Manitius, *Geschichte der lateinischen Literatur des Mittelalters* (Munich, C. H. Beck, 1931), III, 809.

4 *Ibid.;* also Thomas Wright (ed.), *The Anglo-Latin Satirical Poets and Epigrammatists of the Twelfth Century,* Rolls Series 59 (London, Longman & Co., and Trübner & Co., 1872), I, 155.

5 This seems to rest solely on the authority of Leland, *Collectanea,* III, 8, and *Scriptores,* I, 228 (Cf. Manitius, *op. cit.,* p. 810). J. A. Herbert, *Dictionary of National Biography* (London, Smith, Elder & Co., 1895), XLI, 62, says: "Leland calls him precentor of Canterbury; but there is no precentor named Nigel

of the twelfth century,[6] Nigellus speaks of himself, in a spirit of true Christian humility, as "brother Nigellus, least of the brothers of the church at Canterbury, a monk in dress, a sinner in life, a priest in rank."[7]

His first name Nigellus, although it appeared erroneously as Vigellus in all the early printed editions,[8] presents no problem, for it has sound manuscript authority. The surname Wireker, on the other hand, is highly suspect, for it rests solely on the authority of Bale, a sixteenth-century churchman, dramatist, and bibliophile, who calls the poet Nigellus de Werekere, on what authority no one seems to know. It has been suggested that Bale may have misread the name "Wetekere" or "Wetekre," for the name of the village of Whitacre in Kent, which, in that case, would presumably be the poet's native town.[9] There is somewhat greater probability, it seems, that Nigellus' surname may really have been Longchamp, for it does appear as de Longo Campo in a thirteenth-century manuscript containing several of his poems, and is indeed the earliest evidence for his surname that has come down to us.

Assuming that Longchamp (or de Longchamps) is his true surname, some have thought that he might therefore be related to William Longchamp, whom he seems to be addressing in the first verse of his *Speculum stultorum* and to whom he dedicated his *Contra curiales* as well as one of his minor poems. Except for the mere similarity in names, however, I can find no evidence to support any family relationship between these two men. There certainly seems to be no hint of it in the writings of Nigellus himself.[10] But

---

in the extant obituaries of the priory, although the entry, 'Nigellus, sacerdos et monachus,' occurs three times."

[6] Herbert, *loc. cit.*

[7] Wright, *op. cit.*, p. 153: "Cantuariae ecclesiae fratrum minimus frater Nigellus, veste monachus, vita peccator, gradu presbyter."

[8] Thomas Wright, *Bibliographia Britannica Literaria* (London, J. W. Parker, 1846), II, 354.

[9] J. H. Mozley, "Nigel Wireker or Wetekre?" *Modern Language Review,* XXVII (July, 1932), 314–17.

[10] Mozley, *ibid.*, attempts to establish some connection between Nigellus and

this would not, I think, preclude the possibility that his real name may have been Longchamp. While the name Wireker will no doubt persist, it is interesting to note that the more authentic Longchamp (or Longchamps) is gaining acceptance and has already appeared in recent articles on this writer.[11]

As a monk and priest at Christ Church, and as a writer of note, Nigellus must have come into close contact with many of the leading men of his day, although he mentions comparatively few of them in his works. He does, however, speak of Thomas Becket, whom, he says, "I saw with my own eyes, touched with my hands, and with whom I ate and drank."[12] He was on terms of intimate friendship with William Longchamp, as was mentioned above. Nigellus seems to have had a genuine affection for this prominent man and to have held him in the highest esteem, and would seem to be referring to him as "the half and better portion of my life."[13] This relationship, however, close though it seems to have been, did not prevent Nigellus from warning the bishop, along with other ecclesiastics, of the dangers which arise from the neglect of the things of God for secular pursuits. Indeed, the main theme of the *Contra curiales* is to prove, it has been said, "the incompatibility of the office of chancellor with that of bishop."[14] His tone, however, is gentle and conciliatory, although there is good reason to believe that he might have dealt more sternly with the archbishop and

---

the Norman Longchamp family, but his suggestions are hardly supported by the evidence. "Generally," he concludes, "knowledge of the author in the Middle Ages as opposed to that of his work, seems to have died out; no references to him are known."

[11] See R. R. Raymo, "Gower's *Vox Clamantis* and the *Speculum Stultorum*," *Modern Language Notes*, LXX, 5 (May, 1955), 315–20; also his "The Parlement of Foules 309–15," *Modern Language Notes*, LXXI, 3 (March, 1956), 159–60; A. Boutemy, "Une vie inédite de Paul de Thèbes par Nigellus de Longchamps," *Revue Belge de Philologie et d'histoire*," X (1931), 931–63.

[12] Wright, *Anglo-Latin Satirical Poets*, I, 155; *Contra curiales*: "beatum dico martyrem Thomas, quem vidimus oculis nostris, et manus nostrae contrectaverunt, cum quo manducavimus et bibimus."

[13] Prologue of the *Speculum stultorum*, l. 4.

[14] Herbert, *op. cit.*, p. 63.

chancellor, had he felt inclined to do so. As to the real character of William Longchamp the evidence is rather conflicting, but if we can trust what would seem to be a reasoned evaluation of his personality, he was a *novus homo*, physically unattractive, but strong of mind, ambitious, self-confident, and all too ready to achieve his ends even by unscrupulous and intolerant means.[15] There can be little doubt that there was much that he could learn from the preachments of his more tolerant friend and critic, Nigellus.

Most of Nigellus Wireker's active life must have been spent at Canterbury, although it is known that he spent some time at Coventry and was also in Normandy. He speaks of himself as being present at Coventry after the expulsion of the monks from the cloister, and of seeing with his own eyes acts of immorality and vice which it grieved him to relate.[16] Again, he gives a report of a story which he had heard when he was in Normandy.[17] These seem to be the only places beyond the confines of Christ Church which he is known to have visited. Manitius, however, believes that he may have also visited Paris, and perhaps traveled in Italy, because he seems to know these places well, and his works give a general impression of cosmopolitan background. As with his date of birth, his time of death is not known, but it is assumed that he died near the close of the twelfth century, possibly about 1200.

The Works of Nigellus Wireker

The literary production for which Nigellus Wireker is best known is his *Speculum stultorum,* an elegiac poem of over 1900

[15] See William Stubbs, *Historical Introductions to the Rolls Series* (London, Longmans, Green, and Co., 1902), pp. 214–18. Cf. also Mary Bateson, *Mediaeval England* (New York, G. P. Putnam's Sons, 1904), pp. 178 ff.

[16] *Contra curiales,* pp. 211–12. Manitius, *op. cit.,* p. 810, says that this was in the year 1191.

[17] *Contra curiales,* p. 203.

couplets. This was published eighty-seven years ago by Thomas Wright in the Rolls Series.[18] Appearing in the same volume are two other works, his *Contra curiales et officiales clericos*, a prose treatise with a verse introduction, of some eighty pages, and a poem of 153 elegiac couplets addressed to William of Ely, better known as William Longchamp. In addition to these there are a number of still unpublished poems which have been listed and briefly discussed in an excellent study by the English scholar, J. H. Mozley.[19] These poems are preserved in a thirteenth-century manuscript which contains two other poems written in two different hands and which, judging from their style, Professor Mozley regards as not by Nigellus. The others whose authenticity of authorship he accepts are several shorter poems on the following miscellaneous subjects: (1) request to the reader to pray for the eternal repose of the writer, (2) Honorius, (3) the pastoral staff, (4) the duty of monks to abandon all private property, (5) the value which men place on private property and the fickleness of fortune, (6) St. Katherine of Alexandria, (7) the death of Honorius, (8) the good monk, (9) the death of Emma, (10) some reflections on giving, (11) lament on the decline of the world and on the death of Honorius, (12) Honorius, and (13) St. Thomas; a collection of seventeen miracles, under the title of *Miracula sancte Dei genitricis Virginis Marie*; a long poem of 2345 rhyming hexameters, titled *Passio Sancti Laurenti martyris*; and the *Vita Pauli primi eremitae*, a poem of about 700 lines, also in rhyming hexameters.

Historical Setting of the *Speculum stultorum*

The *Speculum stultorum* is a satirical poem, directed against the evils of the day, particularly as they were found in the church in the

[18] *Op. cit.*

[19] "The Unprinted Poems of Nigel Wireker," *Speculum*, VII (1932), 398–423.

late twelfth century. This was a time when much sentiment hostile to the monastic orders had sprung up, and rivalries had developed among the various orders themselves. There had come into prominence a large body of secular clergy who had been educated in the schools abroad, and whose interests lay in the fields of law and government, or in letters, rather than in the contemplative life of the church. Knowles,[20] in his scholarly and most informative book on monasticism in England, says that toward the end of the reign of Henry II a spirit of polished satire and fluent criticism wholly without parallel in previous decades made itself felt among this class. This spirit was not confined to secular clerks; it infused the inhabitants of the cloister too, he adds.

Canterbury, where Nigellus served as monk, was a center of learning and a gathering-place for scholars and ecclesiastics from many parts of Europe.[21] Indeed Canterbury had been a center of monastic life and educational influence in England for many years after the coming of Augustine in 597. In the general decline of monasticism in England, which is said to have reached its lowest ebb by the time of Alfred the Great, Canterbury too suffered, but perhaps to a less degree than the other places. But this decadent state was not to last. The century before the Norman Conquest saw in England the growth of many new monasteries, which attained great vitality and showed no trace of that serious moral decadence and secularization which had earlier characterized monastic life in England, and on the continent as well.[22] Canterbury was one of the strongest religious centers in England. The Norman Conquest itself strengthened monasticism in England, and advanced the cause of learning and culture. Not only did new life and vigor come into the monasteries already existing, but new houses were soon to be established. The period of Henry I (1100–1135) is said to have been

[20] David Knowles, *The Monastic Order in England* (Cambridge, Cambridge University Press, 1949), p. 315.

[21] F. J. E. Raby, *A History of Secular Latin Poetry in the Middle Ages,* 2nd ed. (Oxford, Oxford University Press, 1957), II, 91.

[22] Knowles, *op. cit.*, p. 81.

one of external and internal well-being and the time when Norman culture really thrived on English soil.[23] By the time of Henry II, however, deteriorating forces were at work and there was much which reformers such as Nigellus could find to criticize. The chief critics of this society were those highly educated clerks who served the kings and the bishops, or who taught in the various schools. Their training, based on the Latin classics, naturally led them to cast their criticisms in the literary mold of satire or epigram. Numbered among these critics are John of Salisbury, Gerald of Wales, Walter Map, and Nigellus Wireker.

The *Speculum stultorum*

In the prologue to his *Speculum stultorum,* Nigellus explains the title of his poem and tells how he expects to present his message. He believes that criticism should be administered gently and inoffensively, for wounds, he believes, often respond more quickly to soothing medicines than to cauterization.

Among the evils he attempts to criticize are the vanity of excessive ambition among the clergy, their unreasonable quest for worldly knowledge, and the hypocrisy of men in high position. By holding out a mirror which will reflect the errors of fools, he hopes that others may see and correct their own faults.

The means by which Nigellus wishes to convey his message is imitative of the method used by Apuleius in his *Golden Ass.* He introduces into his story, to represent a monk, an ass by the name of Brunellus (or Burnellus). The ass, like so many monks and ecclesiastics, is dissatisfied with his natural limitations. His various efforts to improve himself make the story which Chaucer called "The Booke of Daun Burnel the Asse." His tail is too short to please

[23] Knowles, *ibid.*, Chapter X, "The English Monasteries under Henry I," pp. 172–90.

him and he is very eager to have it lengthened. He consults the physician Galienus, who points out to him the foolishness of his request and urges him to be satisfied with that which God has given him. To illustrate the futility of his request, Galienus tells him the story of Brunetta and Bicornis, two cows whose tails became frozen to the earth on a cold winter night. Bicornis, impatient and fretful and extremely anxious to be free, cut off her tail and swiftly returned to her home. Brunetta, on the other hand, remained calm, and after a few hours the ice melted and she too was set free, without any harm to her body. When summer came Bicornis paid the penalty for her rashness by falling a victim to a horrible death from the attacks of insects against which she was helpless to defend herself without a tail.

This story, however, makes no impression at all on the ass, and Galienus has the stupid creature go to Salerno to purchase some fantastically ridiculous medicines which will presumably be effective in the lengthening of his tail. Among the articles he is to get are marble fat, kite milk, wolf fear, flash of light, lark kisses, and other such impossible things. Naturally Brunellus has no success in finding these articles, though he walks the streets of the city for several days. At last he falls into the clutches of a London swindler who, for a very high price, sells him a number of glass jars which, he claims, hold the very articles he has come to get. Brunellus takes these jars and sets out with a light heart next day for home. Near the city of Lyons he encounters some fierce dogs owned by a Cistercian by the name of Fromundus. Savagely attacking him, they tear off half his tail and cause him to drop his medicines. The jars break into a thousand pieces. The hapless Brunellus curses the monk and threatens to denounce him to the Pope, claiming that he had bought these costly articles according to his orders. He is sure that the Pope will justly punish him. This frightens Fromundus, and he feigns a reconciliation, but only to allow himself time to devise a means of destroying Brunellus. His plan miscarries, however, for Brunellus sees through his hypocrisy and succeeds in doing to

Fromundus exactly what the monk had planned to do to him, namely, to push or kick him into the river Rhone. After drowning the monk the ass then gleefully sings a song of triumph and composes an epitaph for his tomb. But his mood soon changes, and in a spirit of despondency he condemns himself for his folly. Too embarrassed to return home without any tail at all, he resolves to go to the University of Paris and devote himself to ten years of study in the liberal arts. After that he will go to Bologna to study law, and then, if he is still alive, he will take up the study of the Holy Scriptures and the decretals. All this, he thinks, will bring him much fame and glory.

Brunellus therefore sets out toward Paris. Along the way he meets another traveler, a Sicilian by the name of Arnold, with whom he makes friends and to whom he reveals his recent misfortunes. By way of reply Arnold tells him a delightful story about a cock which took vengeance on the son of a priest of Apulia for an injury done to him six years before.

In Paris Brunellus decides to join the English school, because he finds the English very charming people, generous, and given to wine, women, and song, without exceeding, however, proper bounds. But after seven years of study he has learned absolutely nothing except to say "Heehaw," and that he could do even before he entered the university—in fact it was instinctive with him. He reproaches himself for his stupidity and wishes that he had never been born, but at that moment he happens to recall a remarkable dream which he had the night before, and he believes that it means that he is destined to become a bishop. This is a happy thought, and he dwells at some length upon this pleasant prospect. He recognizes the tremendous prestige which this sacred office confers and the opportunities it gives to serve the people. But nothing less than a full bishopric will satisfy him. He scorns the very thought of becoming a mere abbot. Moreover, he will seek this holy office in an honorable way, not through simony. And he will be none other than an exemplary bishop. This high ambition, to become a bishop, however, is soon shattered, for he recalls now an unfortunate experience of his youth which will make it difficult for him to return to his native land, if the incident is still remembered there.

Sadly Brunellus bids his friends adieu and takes his leave of Paris. From a hilltop outside the city he looks back and is shocked to discover that he has already completely forgotten the name of that city where he has just spent seven years. He wonders how he could be so stupid. By a happy coincidence, however, he soon recovers the name from a stranger whom he meets along the road, only to lose it again in a short time, except for the first syllable. But even this small part Brunellus is pleased to retain, for he has finally learned that it is better to possess one part of a thing than to lose the whole.

As Brunellus meditates along the way there comes to his mind the uneasy feeling that death may not be far away, and he feels the need to do something for the salvation of his soul. He resolves to repent of his sins and to take religious vows.

At this point in the *Speculum stultorum* we reach its very heart, for in the pages which follow the reader is introduced to a critical analysis of the various existing monastic orders. He is shown the advantages and the disadvantages of the various houses. Nigellus describes the way of life of the Hospitalers, Cistercians, Benedictines, Grandimontines, Carthusians, Black Canons, Premonstratensians, Secular Canons, Moniales, and the order of Simplingham. His attacks against the Cistercians and the Secular Canons are less restrained than those against the other orders. On the whole his criticisms are expressed in a mild and good-natured satiric style, occasionally Rabelaisian in character. Perhaps the most effective condemnation of the orders is the fact that Brunellus decides against them all and resolves to establish a new order of his own. He will take from each of the others only those things which please him. The humor of the situation lies in the fact that he will choose their luxuries instead of the bare necessities. His monks will wear more comfortable dress, they will be served better food. From the order of nuns he will take a wife, declaring that the bond of matrimony is the oldest of the orders, established by the Lord in Paradise and blessed by him. But before setting up his new order he must secure the confirmation of the Pope. He will therefore proceed to Rome.

Brunellus is now united again with Galienus, who scarcely recognizes him after his long years of wearisome toil and floggings in the school. After bemoaning his wretched lot, the ass assails in no uncertain terms the decadent state of Rome, her thirst for blood, and her insatiate desire for wealth. Nothing is left there, he declares, that is sound. He attacks the tyranny of kings and their lust for power and possessions. He lashes out against the people who assist them in their wicked designs. And yet he regards the bishops as being even more blameworthy. He calls them false prophets and ravenous wolves. He says that they are avaricious, gluttonous, selfish, and given to lavish display and luxury. He criticizes the practice of appointing children to the high office of bishop while they are still in the cradle. He deplores the preoccupation of bishops with such worldly pleasures as hunting and fowling, their boredom, on the other hand, in carrying out the sacred duties of their office. He finds the lesser officers, also, guilty of insincerity and deceit, and fired with ambition and a desire for riches. Of the religious in general he has little to say, except that they are wolves in sheep's clothing, as sly as foxes, ravenous as hawks, and quarrelsome as butting rams.

Brunellus then begins to inquire into the ways of the laity, but suddenly stops for fear he may bring offense. Instead he narrates a personal experience. He tells how once on a summer day he chanced to be resting under a large oak tree, when all of a sudden a large company of chattering birds alighted overhead and began a conference, in language which he could understand. He listens to speeches from three of the number, a raven, a cock, and a falcon. These are filled with moral preachments and general reflections on life.

Again, Brunellus decides that he will devote himself to the religious life, and he invites Galienus to enter the order with him, urging him not to be ashamed to follow the advice of a pupil, for the wise are often confounded by the ignorant, the weak exalted, and the mighty abased. To illustrate this truth he tells a story of the three fates which he had often heard as a child from his mother. Scarcely has he finished the tale when his nose bleeds, a portent, he fears, of evil; and sure enough, for at that very moment his old master, Bernard, turns up, puts a bridle on him, and drives him back to Cremona. The master freely applies the whip, and to discourage the ass from ever running away again, he cuts off his ears. The poor creature now is not only without a tail, but also destitute of his ears, and he must submit henceforth to his master at Cremona.

This is the last of the experiences of Brunellus, and it might well bring us to the end of the book, but Nigellus gives us one more story for good measure. He tells the story of Bernard and Dryanus, a tale of oriental origin, it seems, and one that was popular in the Middle Ages.[24] There is little motivation for the story except that the events narrated concern Bernard the master of Brunellus, and, like the other stories in the *Speculum stultorum,* it is intended to teach a moral lesson. The story is intrinsically interesting and forms a rather pleasing conclusion to the satire.

In brief, the story is as follows. One day while Bernard is in the forest he hears a cry issuing from a deep cave. It is the cry of the wealthy Dryanus, who has fallen into the cave and is trapped there along with an ape, a lion, and a serpent. He promises Bernard lavish gifts, if he will save him. In the rescue which follows it happens that the three animals are saved first, and only after much effort is Dryanus finally brought to safety. The three animals are grateful for their delivery and show their profound gratitude to Bernard in extraordinary ways. Dryanus, on the other hand, in spite of his earlier promises, shows no intention of rewarding his benefactor. But he is brought before the king and given his choice, either of spending three days in the cave with his former wild companions, or of

[24] J. H. Mozley, "On the Text and MSS. of the *Speculum Stultorum,*" *Speculum,* V (1930), 260–61.

dividing his wealth with Bernard. The king's decision pleases everyone except of course the ungracious and miserly Dryanus, who grudgingly surrenders to Bernard half of all his possessions.

At the conclusion of this story the *Speculum stultorum* is quickly brought to an end. The reader is admonished to give good heed to the lessons of the poem and requested to overlook any faults which it may contain, and last of all to commend its author to the graces of the Blessed Virgin.

That the *Speculum stultorum* was a popular work in the centuries immediately after its publication is evident from the large number of manuscripts extant and from references to the work by later writers. Twenty-seven manuscripts are listed, about half of them in England; the others are scattered about in various libraries in other countries of Europe. Except for one thirteenth-century manuscript, about half belong to the fourteenth century, the other half to the fifteenth.[25] Chaucer in his *Nonnes Preestes Tale* (3312–16) refers to the *Speculum stultorum* as "Daun Burnel the Asse." The Fox is telling Chanticleer of having read in this book how a cock took revenge on a young priest's son, who had struck him on the leg, by causing him to lose a benefice. Again, in his *Parlement of Foules* (ll. 309–15), Chaucer draws his inspiration from Nigellus' description of the assembly of birds (Wright, p. 112, ll. 1–6).[26] And Gower, who borrowed so freely from Ovid, Alexander Neckham, as well as from Nigellus,[27] refers specifically to the story of Brunellus twice in his *Vox clamantis* (I, 201–4; IV, 1189–92, 1207–10) and appropriates such a large number of lines and phrases from the *Speculum stultorum* that his indebtedness has been said to amount almost to plagiarism.[28] And finally, Alexander

[25] Mozley, *ibid.*, pp. 251–63.

[26] R. R. Raymo, "The Parlement of Foules 309–15," *Modern Language Notes*, LXXI, 3 (March, 1956), 159–60.

[27] Samuel M. Tucker, *Verse Satire in England* (New York, Columbia University Press, 1908), p. 85.

[28] R. R. Raymo, "Gower's *Vox Clamantis* and the *Speculum Stultorum*," *Modern Language Notes*, LXX, 5 (May, 1955), 315–20.

Barclay's *The Ship of Fools* borrows the fool motif from Nigellus.[29] The influence and popularity of Nigellus seem not to have been too strong after the fifteenth century, although his *Speculum stultorum* was sufficiently liked to be printed in 1662.

The Satire of Nigellus Wireker

Nigellus Wireker belongs to a long line of Latin satirists beginning with Lucilius, who flourished in the latter part of the second century B.C., and including, in the classical period, such masters as Horace, Persius, Juvenal, and Petronius. Satire seems to have been the name given by the Romans to an early type of literature characterized by the miscellaneous nature both of its subject matter and its form. It covered the widest range of subjects and was often a mixture of prose and verse, or of various kinds of verse. It made much use of dialogue, and frequently contained fables and tales. The style was informal and even conversational in tone, lacking the more exalted qualities usually found in other kinds of poetry. Horace, for instance, calls his satires *Sermones,* talks or conversations. Originally the censorious spirit was lacking and it remained for Lucilius and his successors to instill into this literary genre that element with which satire has become identified.

The degree of the censorious element varies from author to author. Horace (65–8 B.C.) exercises restraint in his use of personal attack, and maintains that jest is more effective than bitter attack. The satire of Persius (A.D. 34–62), written in a strained rhetorical style, is based on Stoic philosophy and fails to deal adequately with the problems of real life. Juvenal (A.D. c. 60–c. 130), on the other hand, is the impassioned reformer who is so outraged by the evils of his day that he cannot resist the impulse to write satire. Indignation is his chief driving force, it would seem, and he is

[29] Tucker, *op. cit.*, p. 164.

altogether unsparing in his use of invective. Gilbert Highet declares that he produced "the most bitter and eloquent social satires ever written."[30]

Writers on the subject of satire frequently call attention to the difficulty encountered in attempting to formulate a satisfactory definition of satire as a literary genre. They observe first of all the freedom which writers have taken with this literary form, the wide range of types and styles, the variations in the censorious element from mild reproach and good-humored banter to bitter and envenomed invective. They are aware too of the fact that many forms of literature have elements of satire in them, or may indeed have the satirizing of society as their chief aim, and yet not fall into the category of satire in the strict sense of the word. Such are the satirical novel, drama, dialogue, essay, or the burlesque, the epigram, lampoon, caricature, fable, or beast-epic.

What definition then can be formulated for this literary genre we know as satire which will apply both to the past and the present? Gilbert Highet's definition is perhaps as inclusive as any. He says: "Satire is a continuous piece of verse, or of prose mingled with verse, of considerable size, with great variety of style and subject, but generally characterized by the free use of conversational language, the frequent intrusion of its author's personality, its predilection for wit, humour, and irony, great vividness and concreteness of description, shocking obscenity in theme and language, an improvisatory tone, topical subjects, and the general intention of improving society by exposing its vices and follies."[31]

With this definition before us, let us look briefly at the *Speculum stultorum* and see to what extent it conforms. In the first place it is a continuous piece of verse and of considerable size. Variety is achieved by the insertion of several illustrative stories, some of considerable length. There is much dialogue and conversational lan-

[30] Gilbert Highet, *The Classical Tradition* (New York, Oxford University Press, 1949), p. 303.
[31] *Ibid.*, p. 305.

guage. The author's personality reveals itself in the very presentation of his message. The element of wit and humor is an integral part of the entire work. There is some obscenity, but it can hardly be said to be shocking, at least to the modern reader; and it is certainly not done to excess. It merely takes the form of coarse pleasantry. His description of the peasant girl in the story of the three fates and his discussion of the advantages and disadvantages of wearing breeches will perhaps best illustrate his use of indelicate language. It can also be said that there is an informal and improvisatory tone to the whole poem, and a wealth of topical allusions.

The very title itself, *Speculum stultorum,* Nigellus explains in his Prologue, was selected with the thought that foolish men might be able to observe as in a mirror the follies of others as if they were their own and to correct them. He hopes that men will thus wish to cast out the motes in their own eyes. In other words, it is his aim that the reader, in seeing the foolishness of others writ large, may thus wish to put aside his own lesser faults. Presenting in allegorical form the experiences of Brunellus the ass, the author, as was pointed out before, desires to criticize by way of mild reproach those faults which he feels sure could never be eradicated by strong rebuke, for he believes that good humor is a more effective medicine than sarcasm or bitter invective. Moreover, in using *speculum* in his title, Nigellus is employing a metaphor which became extremely popular in the Middle Ages. The word occurs very frequently in book titles of the twelfth and thirteenth centuries.[32]

[32] For a brief discussion of the mirror as a metaphor in ancient and medieval literature see Ernst Robert Curtius, *European Literature and the Middle Ages,* translated from the German by Willard R. Trask (New York, Pantheon Books Inc., 1953), p. 336. This book was originally published in German as *Europäische Literatur und lateinisches Mittelalter,* by A. Francke AG Verlag, Bern, 1948.

The Translation

It was my intention at first to render the *Speculum stultorum* into English prose, but it soon became apparent that prose would be a most inadequate vehicle for conveying the effect which the original poem makes upon the reader. While the subject matter itself is largely narrative, didactic, and even prosaic in nature, and while there are few touches of poetic fancy or description, the feeling of rapid movement which is experienced in reading these elegiac couplets can be expressed to better advantage in verse than in prose. But more important, I think, is the fact that each couplet is ordinarily one complete sentence; rarely does an accumulation of subordinate clauses or the use of coordinate conjunctions necessitate the continuation of thought into a second or a third couplet. There are of course some exceptions to this practice. To translate these short units or packages of thought into prose would seem to me to produce a choppy and monotonous effect which is less apparent in a metrical translation. It is for this reason that I have indented each alternate line, as is done in the case of elegiac verse, even though I have made no difference in the number of feet to a line. I have used the iambic pentameter, for it is a common English verse form and seems to convey rather satisfactorily to the English reader the movement of the Latin dactylic meter. A metrical translation cannot of course be as literal as one in prose, but it has been my constant aim to render as faithfully as possible the full meaning of the original text, and certainly not to take undue liberties. Sometimes it has been most difficult to compress within the limits of two iambic pentameters the complete thought of a couplet, but patient endeavor has usually, if not always, made it possible. In general then this may be considered a fairly literal translation.

The text used for this translation is that of Thomas Wright.[33] This is not the perfect text by any means, but it is the only printed text easily available. Its inadequacies have been proclaimed by André Boutemy.[34] It has been my good fortune, however, to have available many corrections of Wright's text which have appeared in a number of scholarly articles on the text.[35] And so, although Wright's edition leaves much to be desired, it can be said with some degree of assurance, I think, that this translation is based on a reasonably sound text.

The modern translator of almost any one of the ancient classics has the advantage, or perhaps disadvantage, of doing that which has been done before. He can profit from the mistakes of others or he may, if not on his guard, perpetuate the errors of his predecessors. Neither is possible in my case. To my knowledge the *Speculum stultorum* has never before been translated into the English language, nor into any other language. I have had to depend therefore upon my own judgment alone in the interpretation of difficult passages. It is my hope that the infelicities in translation will be few indeed, and that the reader will find this version of the *Mirror of Fools* an informative and delightfully entertaining satirical work.

G. W. REGENOS

[33] *Op. cit.*

[34] "The Manuscript Tradition of the *Speculum Stultorum*," *Speculum*, VIII (1933), 510–11.

[35] J. H. Mozley, "On the Text of the *Speculum Stultorum*," *Speculum*, IV (1929), 430–42; "On the Text and MSS. of the *Speculum Stultorum*," *Speculum*, V (1930), 251–63; Walter B. Sedgwick, "The Textual Criticism of Mediaeval Latin Poets," *Speculum*, V (1930), 288–305.

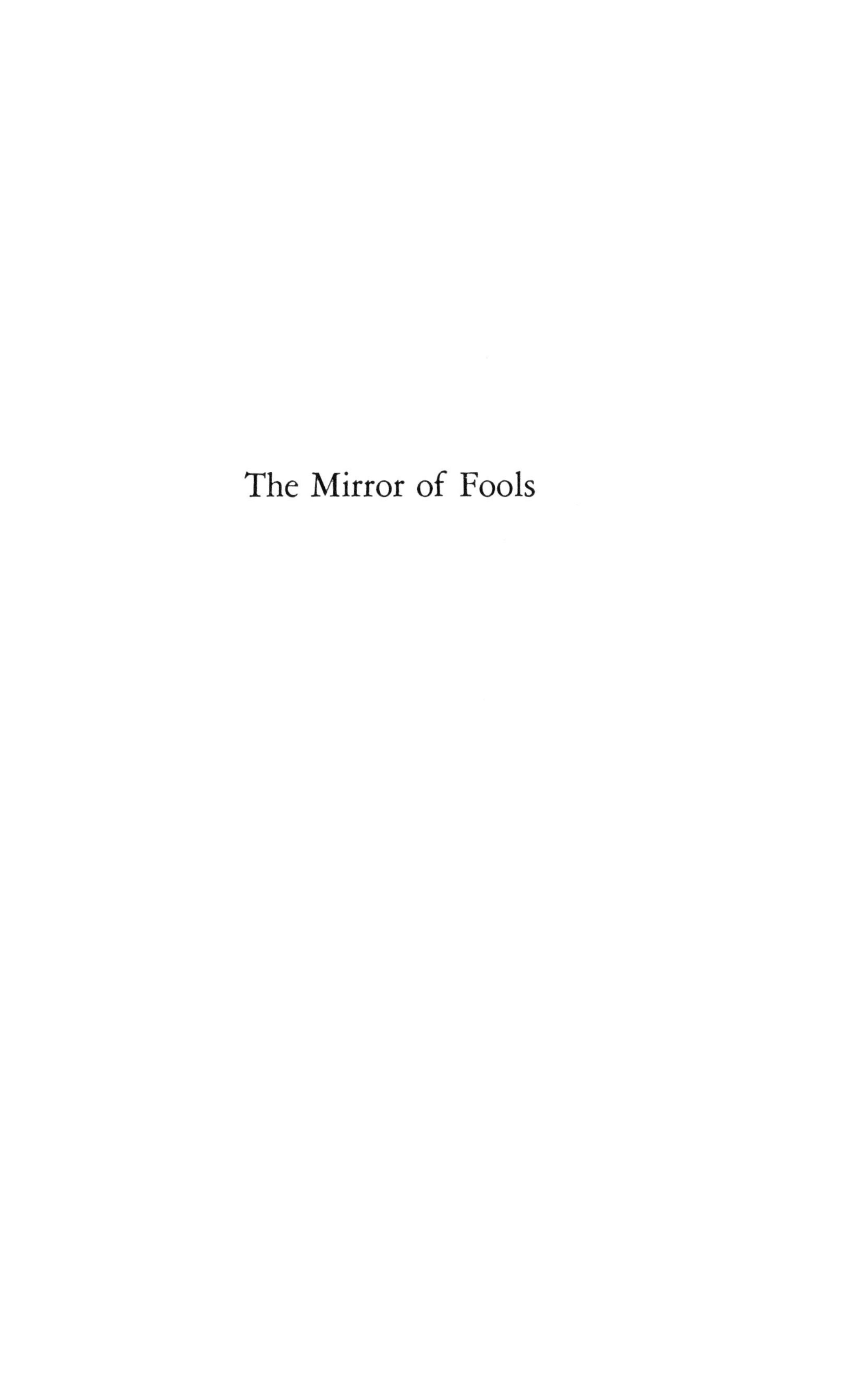

The Mirror of Fools

THE MIRROR OF FOOLS

By Nigellus Wireker

Nigellus' Prologue to the Book Called *Brunellus* Which Follows[1]

NIGELLUS sends his hearty salutations and sincere greetings to William, his ever beloved and esteemed brother in Christ.

I have recently sent you a book, the title and subject of which will seem amusing to its readers and to those who lack understanding. But if you will carefully examine each part and give your attention to the writer's aim and purpose, though the style may be no less crude than the subject matter, you will perhaps be able to gain some instruction from it. That those things, moreover, which are veiled in obscure language, may be clear to you, I consider it proper to explain briefly a few of my hidden meanings. The title of this book is *Speculum stultorum.* It has been given this name in order that foolish men may observe as in a mirror the foolishness of others and may then correct their own folly, and that they may learn to censure in themselves those things which they find reprehensible in others. But even as a mirror reflects only the outward appearance and the form of those who look into it, but never holds the memory of a past image, so it is with fools. Seldom, and then only with difficulty, are they drawn away from their folly, no matter how much they may have been taught by the foolishness of others.

This book is therefore called *Speculum stultorum,* either because fools who have had a glimpse of wisdom immediately lose sight of it, or because the wise profit from the fact that when they have

[1] This title, according to Wright, is taken from MS B.

looked upon the stupidity of the foolish they set their own house in order, and by seeing the beam that is in another's eye they cast out the mote from their own. So much for the title.

Now there is introduced into the story an ass, a stupid creature indeed, who desires to have a new tail grafted upon him, longer than the one which nature gave to him and contrary to nature. This he wishes without any regard to his needs, prompted rather by a silly whim. His eagerness makes him impatient of delay, and stirs him all the more with the desire for that which is forbidden and impossible.

This ass represents a monk or any religious person who lives in a monastery and who, like the ass, is obliged to bear burdens in the service of the Lord. He is dissatisfied with his lot, as the ass with his tail, and toils and strives in every way to secure that which he has not received from nature, and yet cannot receive because nature opposes. He consults a doctor as that one whom he believes able to give him that which in his folly he regards as possible. He wants his old tail taken off and a new one added, for he is utterly weary of the cloistered life, where he must persevere to the end to gain salvation. He strives in every way to be taken off and transferred to a place where he can grow a long new tail, that is, secure for himself a priory or an abbacy and be able thereby to have around him a large following of relatives. These he will proudly trail behind him wherever he goes as a kind of tail.

The doctor, on the other hand, whose desire it is to heal the sick, provided they can be healed, listens to the ass's request, his wish, and aims, but finds his request silly, his wish foolish, and his aims altogether senseless. And so he urges him to give up this idea, to be satisfied with the gifts of nature, and not to covet the blessings of others. He tells him that one can easily lose the blessings one has, but not easily, if at all, recover those which have been lost. And to impress this upon his mind, he tells him this story about the two heifers. Once upon a time, he relates, a certain farmer had two heifers that remained out on a muddy field on a cold winter night. There was a hard freeze, and their mud-soaked tails froze fast to the wet ground. So frozen were they to the ice that on the following morning, although they tried to return to their quarters, they could not break loose, but were forced to stay in that place to which they had so willingly come. One of them, however, being most impatient and anxious to return to her little one left at home, quickly cut off her tail and ran for home, but not until she had urged her companion to cut off her tail too and thus be free to accompany her. The other one, however, was wiser and considered the advantages which a tail that is troublesome in the winter brings in the summer, and she refused to heed her advice. She waited until the noonday sun caused the ice to melt and allowed her to go happily on her way safe and sound, for she was wise enough to exercise good judgment in time of difficulty.

These two young cows represent two classes of monks. There are some who are extremely rash and heedless. When these are overtaken by some adversity, as the heifers caught in the ice, or influenced and blinded by love of family, as the one for her calf, they cut off their tails. They cast far away from their minds, as it were, the thought of death, which should prompt them to guard against the flies that swarm about them and sting. Yet because they neither look to the future nor consider the consequences of their impetuous acts, they cut off completely and irrevocably their old friends who might help them in their hour of need. Consequently, they quite naturally falter when trouble faces them, and are destitute of the help of friends, because they did not wish to put any restraint to their folly.

There are others who are ruled by a more mature judgment. These look ahead wisely and intelligently, and prove their superior wisdom by observing that unexpected things may come to harm them to a greater or less degree. They hold before their mind's eye this saying of the wise man,

"From sunset not the sunrise mark the day,"[2]

watching not the scorpion's head, but his tail. These are the men who are tolerant of the impulsive acts of others in time of adversity; they are slow to remember those who are ungrateful of a kindness, and they look to the end of things rather than to the beginning. These are not unmindful of the summer heat, and they preserve their tails to keep off the flies. The tail is the extreme part of the body, but it is essential to the other members on a hot day. The tail of the body represents the end of our present life, and he who is righteous and guards himself carefully against the flies, that is the snares of the devil, will not be afraid in the day of raging heat, rather of wrath, that is the judgment day. But he, on the other hand, who in the winter, the cool period of his life, cuts off his tail, or rather fails to correct his final deeds, will lie in torment with the devil. Galienus, indeed, wishing to see the ass show wisdom and intelligence, and to look forward to the end, tried to dissuade him, but discovered that it is a very difficult thing to persuade or to dissuade a fool in anything, if he is unwilling. And when he realized that the ass was determined in his desires, he followed the advice of Solomon, "Answer a fool according to his folly,"[3] and advised him to go in search of certain unobtainable objects. It was as if he were plainly telling him, "You do not listen to advice, you do not learn from example; you must learn from the toil of your hands." Indeed adversity has brought wisdom to many who could never be taught by prosperity.

Now the ass who goes away to purchase worthless trinkets, and who conceals them in glass jars, represents a monk who, to accom-

[2] "A casu describe diem, non solis ab ortu." The source of this proverb I have not been able to discover. There are, however, many variations on this theme in literature. The Greek tragic poets warn that 'no man should be called happy this side of death,' a popular belief first ascribed to Solon. We may also compare the maxim, "The evening crowns the day," quoted in Smith and Heseltine, *The Oxford Dictionary of English Proverbs*, p. 434.

[3] Proverbs, 26.5.

plish his vain desires, runs to and fro seeking the favor of men, a thing which is as fragile as glass, and enticing them now with flattery and then with bribes. So many are the difficulties and hardships he vainly suffers that he might have gained the kingdom of heaven far more cheaply. It therefore often happens that if a man goes too far in his quest for foolish and vain desires he may even lose the little which he already has. Thus the ass in his effort to have a new tail lost the one he had through his complete stupidity.

In like manner Brother Fromundus, who sent his dogs against the ass and then pretended when frightened by threats that he would make full compensation, although at that very moment he was devising a clever trick by which to destroy the ass, was himself thrown into the Rhone and drowned. He thus signifies those who strive by clever means to cheat and deceive their less astute brothers, but often fall into the very traps they set for others, according to the words:

"No juster law is there than this,
That knaves shall perish by their own device."[4]

The words which the ass addresses to himself, sometimes foolish, and sometimes wise, depending on his state of mind, signify the inane and foolish meditations of the heart of those who store in their minds impossible thoughts and who look forward expectantly to those things which do not exist, as if they did. They spend their time on unimportant matters, being slothful and careless about important things, but eager and zealous with respect to the unimportant. The works which come after them reveal the foolishness and fickleness of their minds, according to the Scripture: "By their fruits ye shall know them."[5]

The story that comes next, concerning the priest's son and the little chicken which later retaliated because of his broken leg, re-

[4] These verses are from Ovid *Ars amatoria* 1.1. 655–56: "Neque enim lex aequior ulla est, / Quam necis artifices arte perire sua."

[5] Matthew, 7.20.

quires no explanation, for its meaning is clear. Indeed it is a very normal trait of character for those who are injured or offended in childhood, even though the wrong be slight, to hold in their minds thoughts of revenge even to an advanced old age, nor do they ever forget a wrong done to them until it has been fully satisfied by punishment.

Likewise the ass who attends the schools of Paris, but only wastes his time and money on the study of literature (for when he had left the city in which he had spent seven years he could not recall its name), signifies those who do not have minds trained or even capable of learning the letters of the alphabet, and yet to show their learning in both the Old and the New Testaments, read books which they do not understand and forget their titles as soon as they have put them aside. They wish to be considered well educated by their diligent perusal of books rather than to know anything in the true sense of the word. These men seem to be educated until they betray themselves by their speech; and their learning, long hidden, appears as it really is. And yet it is most disgraceful and ridiculous for anyone not to blush to admit that he knows nothing at all when the truth of the matter is brought out. But there are many hypocrites today, and there is hardly a profession or order in the entire world where some hypocrisy is not found. Indeed there are those in the professions who make a pretense of knowledge which they do not possess. And those in the religious orders try to appear other than what they are. Brunellus, then, in pointing out certain practices in the various orders, desires to criticize in a jesting manner those things which he knew could never be eradicated by harsh rebuke. For it is true that many diseases respond more readily to healing medicines than to cauterization.

Prologue of the Author

RECEIVE, O William, these few lines composed
By your old Nigel, in a simple style.
This novel piece I send to you to read,
O half and better portion of my life.[6]
Though foolish on the surface it may seem,
The subject dull, the words not well expressed,[7]
Yet much the careful reader here can find
That's worthy of his study and his thought.
Not just the sound of words must be observed,
But rather than the words, the sense they make.
From simple illustrations who'd deny
That many lessons can at times be gleaned?
A little story often holds great truths,
And in a paltry thing a gem lies hid.
A work which offers aught at all to teach
A moral lesson ought to have a place.
I often wonder as I scan the past
How it by you and me should be observed.
Nought with the past the present seems to hold,
And all is in a topsy-turvy state.
Day comes from night; from daylight darkness comes,
From fool the sage, and nothing from the sage.

[6] Cf. Horace *Carmina* 1.3.8: "et serves animae dimidium meae"; *ibid.*, 2.17.5,6: "A! te meae si partem animae rapit / maturior vis."

[7] Nigellus writes in an equally modest way of his style near the beginning of his *Contra curiales* (Wright, pp. 156–57).

A Cato dull of mind becomes; glib-tongued
Ulysses dumb, as also shift the winds.
A wiser one than Cato and more glib
Than was Ulysses soon will be a mute.
And who was wise in time of cruel war
Is taken for a fool in time of peace
And he who had no wit in time of stress
Becomes a Cato after peace returns.
And that which makes me wonder even more,
Those still unborn rob Nestor of his years.[8]
Thus fortune, in the very act of birth,
Creates strange offspring in a wondrous way.
Indeed a brighter child than Cicero
Is born, and wiser than great Cato was.
All things dependent on the lunar globe
He knows, and many secrets often sees.
A man of learning such as you might wish
To fancy or to picture, he will be.
He'll be like those about him and his age,
And he'll be like his own and by them judged.
It's thus a wall is painted white outside,
Yet on the inside filled with grime and dirt.
Thus paint hides dirt, a jewel graces dung,
And thus does gilding coat the rotten wood.
Who in himself is nothing, nor can rise,
Must needs be lifted by another's help.
Yet 'tis absurd for one to boast too much,
When by another one has been advanced.
Though it's a sin to boast of one's own powers,
Yet this I think to be a lesser fault.
But habit often dulls a sense of shame
And makes man bolder in the course of time.

[8] Cf. *Contra curiales*, p. 179: "Iam enim annos Nestoris adimplevit, et ideo iam delirat et ad ineptias committitur pueriles."

An ass may reign and hold the lion's sway,[9]
And judgment give; he'll always be an ass.
And if on lion's throne an ass be set,
He'll be more cruel than the lion was.
Vainglory and excessive pride betray
The ass, though he be decked in lion's skin.
An act of pretense, if not overdone
And well disguised, might long retain its place.
But too impatient with the new, and vexed,
Old age to its destruction headlong leaps.
When man grows weary of the lot he has,
He'll seek another's to forget his own.
Who badly apes the lion hates to lose
What he for good or bad believed a good.
If he through lack of wit incurs great loss,
It's clear he's lost another's, not his own.
That which he had before remains unchanged,
And he has therefore had no loss at all.
Let him who cannot be a lion be
An ass, for heavy is another's load,[10]
And heavier still the longer it is borne;
In varying degrees loads have their weights.
'Twere better not to grasp another's load,
Therefore, than in disgrace not hold one's own.

[9] Cf. Avienus *Fabulae* 5.17,18, "De asino pelle leonis induto": "Forsitan ignotos imitato murmure fallas; / At mihi, qui quondam, semper asellus eris" (Perhaps you may deceive strangers by your imitated sounds, but you'll always be an ass to me, as before.)

[10] "Alterius sarcina pondus habet"; with this compare Maximianus 5.70: "Intermissa minus sarcina pondus habet."

Concerning the Ass That Wished To Have a Longer Tail

IT HAPPENED once an ass with ears immense
Desired to have a tail in size to match.
Since with his head his tail could not compare,
He deeply groaned about its brevity.
'Twas not because it failed to suit his needs,
Instead, because it was so very short.
The doctors he consulted, with the thought
That they might work what nature could not do.

Galienus' Reply

HIM GALIEN ANSWERED: "Fool, a two-foot tail
Is ample for an ass, why ask for more?
Your tail is long enough; a longer one
Would drag the ground and gather dirt.
If this one doesn't suit you, while you seek
A tail too long, you may lose some of this.
You must not scorn the gifts of nature,[11] but
Regard among your riches what she gives.
Believe me, that old tail upon your rump
Is better than a brand-new tail would be.
Yet this annoys—you want a better tail;
You'll get a worse one through the surgeon's skills.

[11] Again, Nigellus is imitating Maximianus' "Quod natura negat, reddere nemo potest." However, the thought in our text differs somewhat: "Quod natura dedit non sit tibi vile . . ."

A new tail could not easily be grown,
Not even if the old could be removed.
Our life is brief, pain toilsome, time is short,
Exceeding long is art; the cause unknown.
Most difficult are cures, and oftentimes
The practice of the doctors baffles them.
Real risks remain, the doctor's art is not
What you believe, or people think it is.
Not every time does medicine bring health,
Although the doctor practice it with skill.
The art of healing is mysterious;
It eases pain and causes it alike.
It's known that what helps some may others harm;
Not with the doctor rests the sick man's need.
The doctor can desire, not give good health;
'Tis wrought by Grace Divine; God's servant he.
Physicians often fail and are deceived,
And things that harm are, vice versa, good.
Though he hold strictly to what art requires,
Still things will not turn out as he may wish.
For God alone is healer of the sick,
It's ours to wish, but his the power to do.
Without his help and guidance we are nought,
But he can do all things without our help.
Herbs, medicines, and sundry drugs we use,
But he by word alone makes all things well.
The people call us doctors, just in name,
But God is doctor both in name and deed.
Be sure to keep the tail he gave to you,
And seek not, fool, for anything besides.
Hold on to what you have lest it be lost,
And lest worse things may quickly come to you.
If you will trust me, strive to keep that tail
Which he who gave you ears bestowed on you.

The state in which that doctor left you keep,
Until he comes and changes it himself.
It's better to enjoy the doctor's care
Than cherish great possessions in ill health.
The mind oft ponders many silly thoughts
Which hinder more than lead to its desires.
From your own ears you can be recognized;
How foolish is the thing you wish to gain!
With your two ears three bodies such as yours
Could be concealed with something still to spare.
Such ears no honor but a burden are,
For nothing but a curse they bring their lord.
Suppose you had a tail to match your ears,
And these monstrosities had been conjoined?
One blemish on one body is enough;
A double blemish would be worse by far.
For two defects are more observable;
One blemish can more easily be hid.
Long-Ears is now your name, and rightly so;
You are the jest and butt of market-place.
Suppose your tail grew long just like a pole,
You'd then be drawn by head and tail alike.
With two odd things upon one body joined,
Must that not make the people point you out?
Large things befit the large, but you are small.
A larger tail will not advantage you.
I think it better then to keep the old
Than draft on you one less appropriate.
And yet no tail could thus be grown again;
'Twould be an innovation quite unknown.
To talk is easy, but to do is hard;
A flow of words is swift, but action slow.
We speak with ease, for wind and air assist;
To do, however, is no easy task.

Large numbers speak, yet if no deeds result,
  The voice is like the wind, and wind is light.
All men can talk, but if you give good heed,
  Not everyone can thus join deeds to words.
For these two gifts are his to whom, through whom,
  And under whom all things alike belong.
His power is not inferior to his will,
  And he alone confers his gifts on all.
He makes, and then remakes, creates, and rules
  His workmanship, the same in time and space.
All mortal things depend upon his will;
  'Tis he and he alone who gave you life.
What he has granted you, what nature gave,
  Though you rebel, yet such, O fool, are you.
Why do you not sustain that which you must,
  Conforming to that will you cannot thwart?
Believe me, what you thought could be, cannot;
  What nature does not grant no one can do.

Rebuke of Galienus

"INDEED the stag which bounds o'er mountain heights
  Seeks not nor begs for such, does he, O fool?
If greater ones than you, and nobler too,
  Request not such, then why, O fool, do you?
The hare, or bear, the goat, or doe and stag
  Do not dispute the shortness of their tails.
You're not, are you, a better one than they?
  Yet they have shorter tails on them than you.

Consolation of Galienus

"YOUR BODY'S FINE; let not your tail cause shame;
  For if it's short, you have a noble head.

It's not so short, no more than others are,
That's clear, and so it's long enough for you.
A noble head makes up for lack of tail;
Not seemly is it that they be alike.
For fame of tail Louis, the king of France,[12]
Nor his own bishops can yield place to you.
Hold firmly then to what you have, because
You easily could lose your present state.

A Story by Galienus Intended To Turn the Ass Away from His Foolishness

DO YOU not know what fate befell two cows
I saw in childhood on my father's farm?
Brunetta and Bicornis were their names;
The one was black, the other auburn-haired.
It chanced that on a certain winter's day,
Prevented by the night, they came not home.
They wandered over lowlands deep with mud
In search of pastures for their day's repast.
Though muddy were their tails, they couched themselves
As cows will do upon the muddy ground.
The chill of night soon spread across the land,
And rendered solid all that had been soft.
What had been mud before grew hard as rock;
Like marble were the ponds and lakes and springs.
The rivers held in close embrace their streams,
Nor let them in their usual manner flow.

[12] This is Louis VII, King of France, who died A.D. 1180. Mention of this king indicates that the poem must have been written not later than the date of his death.

Each stream became a bridge, now free to cross,
The way was smooth where'er you wished to go.
The ground held firmly half the heifers' tails;
The muddy earth more tightly held the mud.
The earth about was hard, their tails were stiff;
They could not lift themselves, nor journey home.
They both when morning came desired return,
But frozen to the ground they could not leave.
Though trying hard, despairing of success,
At last they wearily restrained themselves.
Yet did they cry and groan with inward pain,
And sought to bring about a change of lot.
They failed, however, for the mighty force
Allowed them no release by their own strength.

Concerning Bicornis

"BICORNIS, fretting for her crying calf,
Was very anxious that she might return.
For yesterday she'd left her young at home
Inside the stable, lest it follow her.
But since she made no progress and had lost
All hope, with deeper groans she thus did speak:

Bicornis' Plan

" 'ONE HOPE REMAINS, to cut our muddy tails,
For they keep us poor wretches from return.
Why have a tail, or what's it ever brought?
What good or honor does a tail bestow?
Behold, I'm kept a captive by it! Why?
No honor but a burden is my tail.
Suppose with tail intact I could escape,
A muddy tail will ever weigh me down.

A dangling tail's a burden and holds mud;
  There's nothing I can see it brings but woe.
I'd now be free and going home unbound,
  Were it not for this dirty tail of mine.
I think it better to remove a part,
  Lest by that very part I thus should fail.
Attorneys claim it's thus observed by law,
  A case perempted terminates a trial.
'Tis wiser then to lose a part than all,
  The tail, a less important thing than head.
I'd rather have my tail torn off, and leg
  Besides, than close my eyes in sudden death.
Perhaps again like this I'll be kept here,
  And just as now I'm held, be held again.
If now my tail is severed off, escape
  Will safer be for me than in the past.
My youngling, if I die, will perish too
  With me, for she is not yet five days old.
And if I fail to do this for myself,
  It should at least be done on her behalf.
The tail's that member of the body that
  Fierce dogs are always first to bite and tear.
Already now they've almost taken half,
  And woe is me to have the rest still on!
The part the dogs have spared I gladly give
  For her, an act of love that saves my life.
Why longer then delay what I shall do;
  It's well to hurry when in evil ways.'
She spoke, and with a knife cut off her tail,
  Then started off for home in greatest haste,
First placing in her sister's hand the blade,
  That she might also loose the bonds from her.
But she, less rash and wiser, patiently
  Received her words, but would not take the knife.

Brunetta's Reply

"SHE THEN REPLIED: 'O fool, what thing is this?
Forbid that I should do what you advise!
When things go wrong there's need for self-control;
You must not be too fast in time of grief.
In doubtful matters such as fortune brings,
The course of action should not be too fast.
Impulse, I know, is bad and brings no good;[13]
If all it does is wrong, then 'tis not well.
You should not grieve when life is hard and rough,
But rather you need help and good advice.
More calmly act in stress, wise Cato bids,
Let ease of mind not yield to bitter grief.
When life is hard there must not be great haste,
Time heals a wound which reason cannot heal.
If fortune now is hard, it soon will change
And lighter make the burden which we bear.
Time rushes on and masters fall, new slaves
Arise, the crowd disbands, bare space remains.
What swiftly comes soon swiftly goes away;
Success along with hardship often comes.
To wish can't turn aside the blows of fate;
Not one of us can know what lies ahead.
It's better meanwhile not to fear the things
To be than know too well and be upset.
For many often learn from their distress,
And many find prosperity brings harm.
A happy lot makes many less alert,
While grave solicitude instructs the rash.
In good conditions pleasure must be feared,

[13] "Impetus . . . male cuncta ministrat." This is an imitation of Statius *Thebaid* 10.704f.: "cuncta ministrat impetus."

Sad hearts more slowly stumble into sin.
Troy stood intact while caution on both sides
Prevailed, and foe from bitter foe withdrew.
The gift of peace razed walls that war could not,
And laxness caused more harm than did the king.
Success is seldom wont to come alone;
Misfortune comes surrounded by a storm.
If fortune lately brought delights to me,
Why should I scorn to bear these slight rebuffs?
Rebuffs I bear, but not for all the year;
Not yet has even one full day gone by.
Were I confined in cell of king or prince,
The chains of iron would gall me more than mud.
Who knows not mud is softer far than iron?
Iron bonds hold kings themselves and satraps fast.
If earthen bonds weigh less than kings' iron bonds,
My lot is surely lighter than is theirs.
They may be kept in chains for five long years;
I may perhaps be free in three more days.
Soft zephyrs blow, the south wind will return,
And meadows will grow muddy once again.
Then will my fetters quickly melt away,
And this my tail will still be left on me.
I could have quickly cut it off, but then
No doubt regretted it throughout all time.
In equal way one night can set it free,
Just as before one night has bound it fast.
One must not think, when certain things go wrong,
That one is under everlasting blight.
Whom winter harms let summer's favor bless,
And let another's fortune lift one's own.
Thus do the times themselves bring constant change,
That we another's honor may enjoy.
A sweet relationship makes friendly ties;

Ties made with me are very strong indeed.
What one unblessed by fortune fails to have,
Another from his richer store may satisfy.
No creature is so excellent or grand
That it can do without another's help.
The changing times with varied risks contend,
Nor are all partnerships on equal terms.
We've often seen small shrubs wreck speeding carts,
And hold in check swift oxen from their course.
A small machine oft shatters mighty towers,
And with a gentle blow high walls fall down.
A little shower constrains the strongest wind;
The east wind settles waters tossed by storms.
Consider nothing with contempt, for it
Can sometime either help or injure you.
Sometimes a thing is worthless you hold dear,
And that is greater which you value less.
Great dangers little things sometimes withstand;
A little whip will scare the raving dogs.
In cities we sometimes see raging fires
With just a little water soon put out.

In Praise of Her Tail

"'ALTHOUGH MY TAIL REMAINS my hindmost part,
Yet I deem nothing of more use to me.
Though now it burdens me and brings no fame,
In summertime it compensates its faults.
For me one tail is better than two horns,
When flies are bad, and more assistance gives.
It fans the wasps, keeps savage gadflies off,
And lashes flies of overbearing impudence.
My tail's a shield to me, a sword, and axe,
A lance, sling, rock, and club, an arrow, torch.

The tail provides my skin a faithful nurse,
And brushes off the dust that clings to it.
It washes, cleanses, and alone serves all
The members, being last, in service first.
If we note well what glory and what use
Each member to its ladyship has served,
Then it alone is worth more to its head
Than all the rest; more busy too than they.
Alone the tail protects the weaker sex
By hiding their pudenda from full view.
These things my tail provides for various times,
Yet at a certain time it pleases more.
The times indeed I fear must needs be feared,
Nor does one fear in vain the common ills.
For when I've suffered anything, or think
I shall, it seems I speak like this:
"Lo! spring is near; the flies will soon appear,
Though still they're hiding underneath the frost.
When heat of summer rages o'er the fields,
And routs and scatters all the herds and flocks,
The fly appears, a deadly pest, than which
No greater plague or scourge is found on earth.
With bites he soon will harass and assail
The wretched herds, nor will he give them rest.
For leaping lightly through the air as if
On wing, he'll bite and sting and drive them mad."
That day will prove it's good to have a tail,[14]
And winter warns that summertime will come.
The day will come in which, if I err not,
The cow will need the tail she now thus scorns.
That day will cause young cows to heed their tails,

[14] In this and the following verses there seems to be a parody on that famous Medieval Latin hymn, *Dies Irae*, which so powerfully describes the Judgment Day, if indeed the *Dies Irae* was written by the time of Nigellus.

And barren ones and oxen then retired.
That day will make Bicornis heed her tail
And rather be without a head than tail.
That day will make tails prized throughout the world,
Which now are less esteemed than broken jars.
That day of doom and dread throughout the world
Will, like the Judgment Day, be filled with fear.
That day will take account of all our tails,
And justly choose which ones are good or bad.
That day will judge the tails of both of us
To see which one is clean and which is soiled.
This is that day a heifer, if perchance
She could possess two tails, would wish for ten.
This is that day the savage fly will come,
The worst tormentor cows and oxen know.
This is that day the dog-toothed flies will come
To bite and sting the members of the herd.
This is that day the herd might well have feared
Would come, lest it should suffer on that day.
How fortunate for me, poor wretch, if I
Could flee that day which frightens all the herd!
That day I fear more even than my death;
For if I be not clean, I'll not be saved.
For flies will be most irksome and the heat
Intense, nor will there be a place to hide.
Not only will the heat and flies annoy
The herd, but masters too of herds and flocks.
This is that day on which another's tail
Will give no help, nor have the power to harm.
No safety can the herdsman bring the herd,
Nor bull the heifer, nor can like help like.
This is that day on which a tail alone
Will help, or places where no flies can go.
This is that day on which, if we confess

The truth, from its salvation ours will come.
So though I'm fastened tightly by my tail
Yet I shall not be ruined by my pain.
I'd rather suffer five or seven days
(Perhaps the bitter cold will meanwhile leave)
Than burn throughout long summer's raging heat
And let the beastly fly make raw my hide.'

How Brunetta Is Set Free

"WHEN SHE HAD SPOKEN thus and slept a while,
Soft zephyrs blew and weather changed, she saw.
The sun was warm and shining overhead,
Its scattered rays had melted all the ice.
The streams were open, earth was warm and bright,
The frost had quickly vanished from the ground.
The fields and woods resumed their normal state;
All things before imprisoned now were free.
Without delay Brunetta rose and lifted up
Her tail, and speedily set out for home.
Bicornis, seeing her returning safe,
Lamented, beat her breast, and sadly said:

Lament of Bicornis

" 'AH ME! What have I done? What madness this?
Why did I rush posthaste to my downfall?
How very hard the end impatience brings!
How grievous is an action done in haste!
Oh wretched me! Oh wretch forevermore!
For me there's nought but woe for all my days.
Alas for me that to all living things
I'm now the laughing-stock for all my days!
Why was my life not severed with my tail

So that my head and tail might both be joined?
Why was I not first swallowed by the earth,
Or stream, or cast into the fire and burned?
And why did I not hasten to my death
By leaping headlong from a mountain top?
Ah me! Why didn't I throw about my neck
The noose and end both life and my disgrace?
Why could not sudden death have closed my life,
Or that dread illness that is everywhere?
Why fret? One healing for my wound is left,
A sudden death, for it alone frees all.
Delay will free me not; dread summer's near;
The winter's gone, and spring is here again.
Bicornis soon to die will live to warn
The folk, and by her folly teach the fools.
Let those who hasten and are loath to stop
Learn not to speed too fast to their own doom.
Count not a momentary pleasure small;
A thing reveals itself in retrospect.
You must think nothing small, if at some time
The same might by some chance bring ruin to you.
The tail harms my own head, and keeps my horns
From causing fear; while from behind I fear.
Why fret in vain? Yet I instruct all those
Who'll follow me to read these words of mine:
While what we have remains, it's nought at all,
But gone, what was as nought before is great.[15]
My tail meant nothing while attached to me,
But now that it is gone it's highly prized.
Though you may fear to lose what's great, regard
Whate'er you have, though it be small, as great.'
At last Bicornis finished these remarks

[15] "Et est magnum quod fuit ante nihil." Cf. Maximianus 1.222: "et redit ad nihilum, quod fuit ante nihil."

Which had been prompted by her grave concern.

What Happens to Bicornis

"MEANWHILE the fruitful summertime had come
To decorate the fields with early flowers.
It now had clothed the trees with leaves, the earth
With grass, and fashioned flowers with equal skill.
Birds freed from winter's prison had escaped
To pay due tribute to the neighborhood.
The nightingale, to compensate her loss
Of speech, fills all the woods with lovely song.
The swallow and the turtledove appear
At nature's summons, each at time ordained.
The lark, dawn's harbinger, comes with the thrush,
Nor do they change the schedule of their lives.
The cuckoo with his repetitious song
Declares the springtime, as he always does.
A blend of disharmonious sounds is heard,
And through the woods a thousand organs play.
The scent of flowers excels the songs of birds,
Their songs excel the pipes; the flowers, sweet balm.
The wood resounds, the fields are sweet with thyme,
The flowers and fruits a spicy fragrance yield.
The ground is warm beneath the summer sun,
And what was mire before is changed to dust.
The earth breeds fleas, and carcasses breed worms;
The brazen fly goes buzzing through the air.
Heat agitates the herds, the gadflies bite;
The nasty sharp-toothed fly descends on them.
O'er hill and ridge and places far away
The herd is scattered, breaking down their stalls.
The gadfly, stinging hard, assails the herd;
And wasps, unmitigated pests, draw near.

Through glades and woods the shepherds with
their flocks
Range far afield in unabated haste.
The herd, hardpressed, is plagued on every side;
The cattle scatter and the flocks take flight.
Bicornis stays behind Brunetta as
They flee, though less secure, yet more intent.
Brunetta with her tail intact resists
The flies by bravely dealing blow for blow.
She switches now at wasps, then gadflies routs,
She drives away the flies in self-defense.
One leaves, another comes; she runs and they
Pursue, they sting and bite, but she speeds on.
At last the angry mob in circling flight
Attacks Bicornis, poor defenseless wretch.
Poor thing, what could she do? Where turn in her
Distress, left helpless in such evil plight?
What she could do she did, while she was free:
She ran and ran with all her might and main.
It's faster far to go on wing than foot;
The flier's wing excels the runner's foot.
Preparedness brings peace, and he who comes
Unarmed to fight is easy to defeat.
The flies and wasps appeared in single mass,
The savage sharp-toothed gadfly also came.
In headlong flight she sped through all the land,
And hurried on till she was overcome.
No less did shame of tail drive her away
Than those sharp pricks by which she was beset.
At last beneath a mountain ridge, with head
Thrown back, she fell and quickly sank to earth.
And though she now was close to death, with strength
Renewed she held her head erect a while,
And seeing her Brunetta and her friends,

Who mourned her death, she said to them these words:

Lament of Bicornis

" 'HOW SAD THIS DAY, how memorable to those
Who live, can be discerned from my mistakes.
Death brings no grief to me, for should I live,
That life would still be worse for me than death.
For wretches death is sweet;[16] for me relief;
No kinder thing than death to wretches comes.
To die is not a burden but a boon,
For death is that which keeps us safe from harm.
A death unsought brings peace which life denies;
I'm losing nought by death, but find relief.
I've lived to warn the many, and my death
Will teach them all the need for self-control.

Confession of Bicornis

" 'PERCEIVE HOW VERY NEAR the brink of ruin
And senseless is the glory of the world.
Our passions turn aside what fortune gives,
And nothing is secure which it presents.
Perceive how vain are fleeting pleasure's gifts,
Those gifts she falsely claims to be her own.
No special gifts does nature bring, but grants
That all be common, for the greater good.
Perceive the evils which my tail has caused,
That it has been the cause of death to me.'
She wished to say some more, but growing weak,
All she could say was 'Bruna, farewell to thee!'
The herdsmen then to keep her name alive,

[16] Nigellus is following Maximianus here word for word, 1.115, "Dulce mori miseris."

Set up this epitaph to mark her tomb:

Epitaph for Bicornis

" 'THIS ONE, herself a fool who taught the wise,
Has now gained wisdom, but is food for worms.'

Galienus' Conclusion

"THESE THINGS THEN, O Brunellus, which I've seen
Myself I've told that you may learn from them.
Let what you have suffice, because it's true
You ask a foolish thing; your honor keep.
But you must never think this thing can come
From medicine; you'd better stay at home.
If all the doctors of the world should swear
To grant you this, 'twould be so many words.
I do not say this tail of yours cannot
Still grow; but no new tail is possible.
The doctor cares for bruised and broken limbs,
But amputated members so remain.
Dead things with live things can't be joined at all;
What's cut but not cut off admits of cure.
Your tail therefore can reach a proper length,
Provided it keep contact with the flesh.
But in the case of treatments and their costs
It's quite important that you have much means.
Strong remedies require a bulging purse;
Large wounds demand much wealth and capital.

Ironical Advice of Galienus

SUFFICE it not for you if it should grow
A cubit longer than it was before?"
At this the ass let out a little laugh,
Yet such that it was heard throughout the town.
Then Galien said: "Off to Salerno go
At once and fetch the needed medicines.
Make haste and soon come back; don't tarry long,
And give yourself a fifth foot, if you can.
If then you do not fail to pay the bills,
I shall not fail to give you special care.
If you keep up the payments, I shall try
My skill, provided I still have my strength.
By speed therefore and double time you'll make
The journey short, for it is very long.
You're known to many and have loyal friends
Who'll help you and present you lavish gifts.
You'll find these drugs perhaps quite easily,
But still, I think, they'll not be quickly got.
But you'll take with you all the proper jars
In which to bring these things and keep them here.
The greatest care and effort are required
In packing, lest these things should be destroyed.
All things well packed and placed in proper jars
Can be more safely handled and conveyed.
Though you're not dull of mind, mark on your skin
(For you can write) the formula you need.

Prescription Which Galienus Gives Brunellus for the Lengthening of His Tail

THESE ITEMS MARKED with labels you'll bring back
(It's your concern to see they are not lost):
Some marble fat, a sevenfold oven's shade,
The offspring of a pair of mated mules,
A small amount of milk of goose and kite,
A little flash of light, and fear of wolf,
A dram of seven-year truce 'twixt dog and hare,
The kisses which a lark has sent her hawk,
A pound of special peacock's sweet refrain,
Before however he has grown a tail,
Some red unwoven cloth without the warp
(And you of course will give your ass's laugh),
A pound of yellow herring sperm or bee's,
Some liver, blood, or foot of tiny mite,[17]
A small amount of holy Christmas night
(Its greater length will be just right for this).
On coming from the Hill of Jove[18] you'll take

[17] The Latin word is *cirolus* and has good MS authority, but its exact meaning is uncertain. Mozley, *Archivum latinitatis medii aevi*, IX (1935), 92, offers several possibilities, choosing as the most likely the meaning "picarel," a very small fish. Migliorini, *ibid.*, pp. 256–58, discards Mozley's various interpretations in favor of "ciron-worm," or "mite." He believes that the Latin word comes from a latinization of a word of Germanic origin, found not only in the German dialects, but also in the French dialects and in the literary languages from the Middle Ages to the present time. His argument seems convincing from the linguistic point of view, and his meaning of the word satisfies the context in which Nigellus is listing these things which are, for the most part, *impossibilia.*

[18] "Mons Jovis." This is a name which was given to the Great Saint Bernard Pass, on which there stood in Roman times a temple dedicated to Jupiter. Cf. the Anglo-Saxon word *muntgeóf* in Bosworth-Toller, *An Anglo-Saxon Dictionary*, defined as "The Alps."

From off its peak four pounds, less one full ounce.
On Saint John's night[19] amidst the Alpine heights
Bear off from there some newly fallen snow.
And also get some ruddy serpent's tail
(It's very useful, but don't purchase it).
These well selected items, while still fresh,
Upon your shoulders throw in little bags."

Reply of Brunellus

To THESE HIS WORDS the suppliant ass replied,
On bended knee and with his head bowed low:
"All this I'll gladly do; I'll hurry there
And back; my life, no other's is at stake.[20]
Though always slow before, I'll not be slow
In this, if God will grant me all success.

Brunellus Asks Galienus for His Blessing

"LO! I AM OFF! Your blessing as I go,
That all my life and way may have success."
Then Galien, turning to his mother tongue,
And smiling slightly, prayed and blessed him thus:

[19] There are two feasts of Saint John, one on December 27 in honor of Saint John the Evangelist, the other on June 24 in honor of Saint John the Baptist. The midsummer festival was much more important than the one in December and probably the one referred to here.

[20] "Nam mea res agitur." Cf. Horace *Epistulae* 1.18.84: "nam tua res agitur."

Blessing of Galienus on Brunellus

"ALMIGHTY GOD grant you a thousand woes,
And may your tail receive ten thousand woes!
May water be your drink, large thistles food!
May marble be your bed, your roof the dew!
May hail and snow and rain attend your way,
And ice and hoar-frost cover you by night!
And may a savage dog take after you!"
The ass then gave to him a kiss and said
A loud amen which echoed through the town,
And it was heard through all the market-place.

How Brunellus Sets Out to Salerno for the Prescription

IN HASTE he therefore started off, but at
The threshold caught his foot and tripped himself.
"A sign for his return," the neighbors said,
While others laughed, and to himself he spoke:
"A feeble start will lead to better luck,[21]
The road is long but will not so remain.
Bad luck enough I've learned since I was born,
And I can bear quite well the very worst.
This frame of mine is used to evil things,
And not to luxuries or drunkenness.

[21] "Debile principium melior fortuna sequetur." Cf. Ovid *Metamorphoses* 7.518: "flebile principium melior fortuna secuta est."

The burr and thistle are my choicest foods,
And rain provides for me what I shall drink.
My heart's desire is not for pickled fish,
My body thrives all right on simple fare.
I'm slow and lazy, but would surely be
More so, if lavish foods were ever mine.
But it's not good that he who has great wealth
Should always want to drink Falernian wine.
What need have I for wine, by which the wise
Are turned to fools and countless ills befall?
My parents carried wine upon their backs,
Not in their stomachs; far from me be wine!
So I shall carry wines but never drink,
Lest fever come upon me every day.
If I opposing nature chanced to drink,
It soon would give me quartan and the mange.
I have four reasons for avoiding drink,
Although it does make sluggish minds alert:
To save expense, with fever not to burn,
To keep my wits, and not degenerate.
It's clear that he who can, but won't, shun sin
Deserves less pity if he basely yields."

How Brunellus Comes to Salerno

Now after he had traveled twice six days,
He saw the lofty walls toward which he moved.
And soon on bended knee, with arms raised high,
He humbly paid his vows to God and prayed:
"Almighty God, for Saint Julianus' sake,
Thy mercy and a kind reception grant!
Far be all ills as bridges and twin gates!

I find enough of these in my own land.
May not a peasant or his sack be found,
Nor may that town possess a mill, I pray!
May spurs be dull and dogs be deaf and dumb,
And far from me be bark of vicious whelp!
A country host be mine, with full supply
Of prickly thistle weed and water too!
May wares be cheap and buyers few in town,
And may each seller be in need of cash!
May stores be stocked and money very scarce,
May there be rain, may mud lie in the street!
May all go well and may I have success
In order that my trip be quickly made!"
As he thus mused, he came into the town,
He found an inn for rest to weary limbs.
Next day he rose and went down town to shop,
But finding not these goods he sadly left.
He searched four days throughout the city then,
Yet did not find the things he wanted there.

How Brunellus Is Deceived by a Merchant

A LONDON MERCHANT saw him as he made
This search, and seeing who he was thus said:
"A splendid agent you, whoever sent
You here, for you seek great and costly things.
A wise and rich man he who sent you here
Like this to represent him, that is clear.
Indeed your face and mien reveal full well
The place from which you come and who you are.
You're famous, and though dressed in strange attire,

You'll be no stranger in your native land.
Oh, how concerned for you are all your house
And for your safe return make vows and prayers!
An Englishman am I, in London born,
I too am here to represent my lord,
For nature gave to him too blunt a nose,
And he by science aims to cure this fault.
Four men therefore my chief sent to these parts,
Because it was a thing of grave concern.
Year after year has passed since we arrived,
And three companions have been lost by death.
I hold here objects which, if fortune let
Me carry home, would be most troublesome.
What in this city you have come to seek
I have it all sealed up and stored away.
My money has been spent, I've borrowed much,
Nor to my homeland may I now return.
But I am homesick and desire the more
To go back home the longer I'm denied.
I've been informed about my father's death,
And rumor has it that my lord is dead.
I have ten glassware vessels, and they're filled
With all you list and many things besides.
If you will pay the price I paid myself,
Tomorrow you can leave for home in wealth.
Forbid that I a stranger take from you
A stranger more than I have paid myself.
Now when you've taken this (you seem devout),
Permit that I be mentioned in your prayers."
Forthwith Brunellus paid the price and took
From him those ten glass vessels filled with trash.
And that for each the contract be more sound,
He asked the seller's name, and he replied:

"At London Gila gave me birth, and from
My father's name Truffator I am called.[22]
My sister Bula is of great renown,
And Truffa has become my wedded wife.
No less must you yourself regard it right
To tell me who you are that I may know."

Brunellus' Idle Boasting

"BRUNELLUS IS MY NAME," replied the ass,
"Quite widely known by name and deed alike.
My country's princes and the kings themselves
I serve, yet do not lead a servile life.
My father and grandfather always served
The kings and will for all the time to come.
I'm in the king's employ by family right,
And all the court requires my services.
When I am brought before the king, he yields
To me, and I go where he leads the way.
But I have one most disconcerting pest,
A ruthless rustic, such a deadly curse.
For always he bears goads or knotty whips,
Or walks along with dogs on either side.
My father loathed him, and my relatives,
And all my people who have ever lived.
I like the ways of nobles and the prince,
Their seemly laughter, jests, and pleasantries.
The king says 'Hail!' to me each time I pass;
'Fie, fie!' earth's bane, the simple rustic cries.
Let mange assail him, torture him each day

[22] *Truffator* and *Truffa,* two lines below, are Late Latin words; the former meaning "trickster," and the latter, "fraud," "deceit."

That he can neither ride a horse nor walk.
May he choke up with phlegm, and headaches have,
And may he have the cramps both day and night!
A heartless master may he long obey,
And may his wife and he forever quarrel!
May he be plagued from sole of foot to crown
Of head with ulcers filled with worms and pus!
Him I commend to Satan from whose dung
He sprang, as he himself is ample proof.
If such I could destroy through all the world,
There'd be no place for him on all the earth."

How Brunellus Returns

BRUNELLUS then with joyful heart made plans
To set out for his native shores next day.
When morning light appeared Brunellus packed
His bags, made vows and prayers, and left the town.
In times of joy how near disaster lies!
Yet none can clearly see the course it takes.
He soon drew near to Lyons on a path
Which led across the fields, and thought no harm.
But him a certain of the White Monks, near
Whose house he passed, observed some distance off.

How Brunellus Falls among Dogs

FOUR MONSTROUS sharp-toothed hounds he sent on him,
And in the words which follow said to him:
"You can't believe the path across our lands

Is more direct, and wider than the road?
Perhaps you've lost your way, but don't you think
That these four dogs can put you on your course?
The winding path across our fields could not
Surpass the road, or am I just a fool?
I'll show you how direct the road is through
Our fields, unless these four shall fail to act."
He spoke, and loudly urging on his dogs
He chased him off, assailed on every side.
Brunellus was too slow; the dogs too swift.
They dragged him with their teeth, and threw
him down.
Grimaldus first, in most atrocious way,
Caught hold his tail and tore it half away.
And he, without a thought of all the things
He held, and that the vessels were of glass,
On being seized and bitten by the dogs,
Fell headlong, dropping all the things he had.

Brunellus' Misfortune

THE VESSELS all were lost by this one fall;
A slight mishap could break such fragile things.
Things gained by toilsome effort fell with ease
And brought a lasting sorrow to their lord.
Those things alone remained which had no worth,
For not a thing save bits of glass remained.
His substance all had passed into thin air;
The object of his hope and toil was gone.
He then complained for losing half his tail;
He also grieved for having lost his wares.
Both evils vexed him; one was quite enough.
He grieved to bear the weight of double loss.
He knew not what could cause him greater pain—

The toil, expense, or cutting off of tail.
Indeed his toil was vain, his money spent,
And no one could restore his severed tail.
Distress of mind provoked by pain increased.
He suffered pain and shame successively.
But shame when yoked with pain is worse
Than pain alone when it is free from shame.
That wretched creature in the mouths of dogs
Was rolled about with life scarce left in him.
Nor did the hounds release their hold, until
Fromundus with reluctance held them back.
The monk Fromundus, like his brother monks,
Not like a beast, nay long a beast disguised,
As is the manner of those wicked men,
Began at last to amble slowly there.
He went not hastily as was his wont
When called to meals, his stomach leading him,
But with spondaic foot and ass's pace
He came as in the night he goes to lauds.
And stretching forth his hand: "Bless you, ha! ha!"
He said, then caught and drew away the dogs.

Reply of Brunellus

"MAY YOU BE NEVER BLESSED," Brunellus said
To him, "but cursed for all your days to come!
You didn't learn, did you, from Benedict
To send your dogs 'gainst strangers? No indeed!
A new command, a brand new rule is this
That you've been given, brother, to observe!
Not at Cistercium[23] was found such rule,
Nor came from there your order's stern decree?

[23] *Cistercium* is the Latin word for Cîteaux, the site in France where the Cistercian Order was founded in 1098.

The fathers did not come together and
  Establish such a rule this year, did they?
To plague with dogs a servant of the Pope,
  I do not think your rule provides for that!
Nor do I think it's granted you to kill,
  At least not when no cause of death is found.
Conveying potent ointments of the Pope
  I came here with no thought of harm to you.
As guest and stranger I admit I've crossed
  Your fields, yet walked along an open path.
The path was wide (I took it for the road),
  'Twas very wide and quite well-traveled too.
The wrong is one of deed not of the heart,
  If there is any wrong at all it's slight.
A slight offense should not be punished thus;
  The minor crimes deserve more leniency.
Although the wrong deserved some punishment,
  There was no need for punishment like this.
The penalty imposed should fit the crime,
  For one too harsh is in itself a crime.
As agent and a servant of the Pope
  I came in peace, but chanced to lose my way.
Respect and honor for the highest priest
  Should mitigate somewhat your punishment.
But they've increased it, and in all these parts
  There's none who fears the Lord and his high priest.
It's not to me but God grave wrong is done;
  The Pope will guard my honor, also God.
Upon my lord the Pope the wrong redounds;
  By such a wrong he's hurt, and all his court.
This dreadful crime, this very loathsome deed,
  This heavy blow, shall it not be avenged?
If in contempt so open to our lord
  The Pope, Rome uses not stern means to make

Dishonor short for friends, but honor long,
  And honor and dishonor come to foes,
Then there will be no other power to fear
  And each may do by law what each desires.
Upon our lord the Pope dishonor falls,
  And I alone sustain all other wounds.
Two thousand marks or more my loss has been,
  Besides the cost and troubles of the trip.
I know not what disgrace will cost, but he
  Will know whom honor and this loss concern.
The shame is great, but punishment will be
  Inflicted by the Pope, and crime avenged.
The court at Rome which renders good for good
  Will give a good reward for worthy deeds.
The case is great, the person greater, and
  The loss is very great; of weight each one.
This mainly falls upon our lord the Pope;
  He's deeply hurt and has two open wounds.
The greater wrong he'll readily forgive,
  The lesser will bring heavy punishment.
A wound to honor can be healed by tears
  And prayers, lest he appear to be too harsh.
But losses he will not at all forgive;
  He's kind, but he must have the loss redeemed.
While the Cistercians are in solvent state,
  He cannot altogether suffer loss.
Let them pay tithes or by established law,
  Though loath, conceal their thighs with pantaloons.
They shall not leave the cloisters and shall lose
  Their farms, or pay in full three thousand marks.
Three thousand marks will settle their accounts,
  But if they're slow, they'll pay a thousand more.
But as for me, what vengeance shall be mine?
  Just how shall I so mangled get revenge?

The Pope indeed will rule that not a monk
  May go at all beyond the cloister walls.
If anyone shall violate the rule,
  He must observe forever these three things:
Forever drink no wine, but eat raw beans,
  And suffer punishment both long and hard.
The Pope will gladly aid with heaven's laws,
  But what will be performed by laws of earth?
The law I'll make throughout the world is this,
  If life remains for me so long a time:
That all those converts, rather perverts as
  They're known, get certain names to suit their deeds.
Who catches sight of them outside the church
  Let him forthwith their eyes and feet destroy.
And if a bell be hung not from their necks,
  Then let them be emasculated too.
Let them wear breeches, let them have no wine,
  And cursed be flesh and fish and eggs alike.
Let only two cooked foods suffice for all,
  And even raw, except on holidays.
I'll have my kinsmen ratify these rules,
  That they may be enforced forevermore.
And these the Pope will also have confirmed,
  And they will be endorsed by all the court."

The Hypocrisy of Brother Fromundus

WHEN HE HAD HEARD these words Fromundus feared,
  And stupefied beyond belief he said:
"If this the abbot learns, as learn he must,
  The drowning bag and certain death remain.
What madness drove me to abuse with dogs
  The servant of the Pope? I didn't think.
There's need for action now and that quite soon;

There's need for help and counsel for myself.
For it is true if he escapes alive,
In no way can the matter be concealed.
There's need that I cajole him with soft talk,
Until there comes a time to do this deed.
It's right that he be taken from our midst,
Lest he bring total ruin to our house.
But such a monstrous act requires great skill,
And far from people must be carried out.
And if a gossiper learns aught of this,
My troubles will be greater than before.
My brothers more than gossip do I fear,
Because their folly knows not where to stop.
Some limit can at least be set on talk,
But them no one can hold in check when mad.
To tell a secret to the monks is like
Revealing to the town my hopes and fears.
A thousand faces have the monks, the town
Has one; so keep your secrets to yourself.
For trust that lasts a month will change in time.
Trust not a pledge, though mine be ever true.
Who strives that many have his confidence
Can quickly lose that confidence, I think.
'Twill be a secret; none save me alone
Must know the things that shall be done outside.
Why linger? Humbly shall I go and beg
His pardon, promise to make good his loss.
I'll greet the fool, pretend to be his friend,
And he will quickly put his trust in me."
He spoke, then humbly came and pardon begged,
And pledged Brunellus more than he had lost.
Fromundus welcomed him and bade him rest
From endless toils, and on his cot relax.
He promised wealth, physicians, general feasts,

Fresh air, and anything his heart might wish.
He'd lack for nothing, and when once he'd gained
His wishes, then he might survey the grounds.
For close to Lyons near the river Rhone
There is a place that's covered thick with trees,
A second paradise, for what the mind
Of man desires the blessed earth provides.
The master of that place will so arrange
The house to suit his will, no one oppose.
A slave will heed his beck and call; each one
Will want to please him and to do his will.
He shall himself give orders how he wants
These clients to obey his every nod.
Not like a guest or stranger shall he come,
But like a lifelong member of his town.
He begged, abjured, and promised everything,
And pressed his claims by voice and clasp of hand.

The Hypocrisy of Brunellus

BRUNELLUS gave his free assent, but on
Condition that he promptly kill the dogs.
He understood Fromundus' guile and tried
To hide his thoughts beneath a gentle smile.
He said a few words to himself but low
Enough as not to reach the stranger's ear:
"Behold an ass's thoughts and threats conflict;
Discordant wishes various ends obtain.
The good is often swifter than the bad,
And skillful care forestalls a bad disease.
Deceit yields to deceit, and craft to craft,
And guile is overcome by swifter guile."
Fromundus then with club attacked the hounds
And sought to trick his comrade by a ruse.

And while along the river Rhone the ass
Advanced, and close behind Fromundus came,
Brunellus, seeing what a splendid place
That was for fatal leaps and sudden death,
Cast down Fromundus from the rocky height
And hurled him to the bottom of the Rhone.
Engulfed, Fromundus tasted death, and by
His own deceits was first to be deceived.
Brunellus then burst out in boisterous song,
And loudly sang this jubilant refrain:

The Song of Brunellus[24]

"LET'S SING, O ASSES! Let us celebrate!
Let's stretch our vocal chords, my friends, in song!
Let asses dance and sing a joyful song,
Let cymbals, drums, and choirs ring out in praise!
Grimaldus perished, and his comrades too;
One death has taken my four enemies.
Fromundus fell into the Rhone and drowned,
He suddenly was tricked by his own schemes.
Fromundus fell into the deadly trap
Which he unknowing set for his own self.
The stupid fool whom he contrived to kill
Anticipated him, for he was wise.
Into the pit Fromundus headlong fell,
Which he before had dug with his own hands.
Now lest the journey be too long, the monk
Traversed it with one leap quite unaware.
The leap was unexpected, steep the rock,
The current swift, the river very deep.
He made a dive which he had never learned;
No other did he need, for one sufficed.

[24] This song of Brunellus seems to parody the hymn of praise, or canticle.

Exult, O Rhone, in triumph's wide acclaim!
Fromundus is no more; let earth rejoice!"

Burial of Brother Fromundus

FROMUNDUS, having drowned, was taken from
The stream and buried in his native soil.
The ass beheld his tomb as he passed by
And wept, then carved in stone this epitaph:

Epitaph for Brother Fromundus

"FROMUNDUS, warning to his brother monks,
His living foe has thus immortalized:
He swift and wise would trick the slow and dull,
Whereas the latter came and tricked him first.
The slow and dull thus caused the swift and wise
To be a fool by swiftly leaping down.
Guile conquers guile and art is tricked by art,[25]
Thus fraud and guile bring forth their due rewards.
Let monks recall Fromundus' fatal leap
And learn a lesson from this evil deed."
Brunellus then, when this was done, made haste
Toward home, alert in body, mind, and foot.
But being very poor and short of funds
He often stopped to beg along the way.
He often called to mind his own mishaps
And pondered Galien's recent words to him.
The more he weighed his turn of fate the more
He blamed himself and thus addressed himself:

[25] Nigellus' wording here is precisely that found in Cato *Distichs* 1.26.2: "sic ars deluditur arte."

Brunellus Rebukes Himself for His Laziness and Stupidity

I'M NOT the wise Brunellus, but a dolt,
An ass forever, prince of fools, a dunce.
A fool I was when born, a fool before,
And nothing but a fool shall always be.
A fool my father was, a perfect fool
My mother; nature's made a fool of me.
What nature gave, what's fixed by course of time,
The same abides, holds fast, and is unchanged.
Alas! What sort of shameful thing I am!
Though I think hard, I know not what I am.
I've spent my money and myself as well,
I've also wasted time and effort too.
I'm now quite old; thus far, I must confess,
I've been a fool—the proof of it is clear.
Dull minds are always shameful in the old;
Unless the mind is keen, old age brings shame.
Senility brings laughter to the young,
Nor do the passing years restore their loss.
However wise the old man is, still youth
Will always think the old man is a dolt.
And if one be both old and weak in mind,
The young will judge him nothing but a fool.
A trait of many elders is to think
Themselves more wise the stupider they are.
My master's words and looks revealed to me
How far from reason were my heart's desires.
Alas! How often, if there'd been a way,
He would have kept me from the things that harm.

But fruitless folly and a blinded mind
  Led me to scorn and jeer at what was good.
And what will happen in the morning when
  I come back to my city quite unchanged?
A blemish on the rear is sometimes worse
  Than one in front—I'm fearful for my rear.
Will not the sheriff and the people see
  Me coming back and my misfortunes learn?
They'll say of me: 'As when he left the town,
  Behold, he's back again with nothing new!'
That could indeed be said, if only, but
  It's not, my tail were as before intact.
What then will be their talk when they behold
  This dreadful loss? Can they hold back a laugh?
I'll be the jest of all the citizens,
  My stubby tail the jest of all the town.
If heretofore Brunellus was not on
  Folk's lips, he will be then, and pointed out.
And whom the cloister's rules forbid to speak,
  In silence they'll point to my stubby tail.
And so the talking finger in return
  For curb on speech, will point to all my shame.
If from so many chances you could find,
  Brunellus, one, how great would you become!

Brunellus Decides To Study before Returning to His Family

IT'S BETTER that I not return home maimed,
  Not till I hide my faults by change of lot.
My body still is sound and fit for work;
  I'm not, unless I'm wrong, a hundred yet.
I'm still alert and strong, and have good health,

I need this sluggish mind made sharp by school.
My mind and body both can well endure
Long hours of study and of sleepless nights.
I still have many years to pass before
I reach an age as great as was my sire's.
I'll not, like boys, be hurt by heavy whips;
From youth I've learned to suffer many blows.
The wanderlust won't dull my zeal for school
Which from my seriousness will grow and grow.
The weight of years will banish levity,
And strict routine alleviate the load.
Nor will respect for years, since what I take
Will be for children, keep me from the schools.
No fear or disappointment shall prevent
My pressing toward my goal both day and night.
Since I'll be brave, hard work will conquer all,[26]
And God himself is wont to aid the brave.
I'll make my way to Paris to devote
Ten years to liberal arts—I'll not delay.
Then to Bologna I'll return, if God
Directs, to gain a knowledge of the law.
The Holy Scriptures and decretals shall
Complete my efforts, if my life is spared.
And so I'll have the title "Master" placed
Before Brunellus, and I'll thus be called.
If one should say Brunellus, but perchance
Leave "Master" off, he'll be my public foe.
If thus my famous title goes before,
I'll be a public speaker without peer.
The Senate and the people will rush forth,
The rank and file will cry, 'The master's here!'
The bishop and the monks will all agree

[26] "Labor improbus omnia vincet." Cf. Vergil *Georgics* 1.145f.: "labor omnia vicit improbus."

To rule themselves by my advice and help.
What I establish in the city shall
Remain, my words shall be as good as law.
What then I lack in tail but have too much
In ears, my famous name will compensate.
I'll gain far more than I have lost,
My final glory will blot out all faults."

How Brunellus on His Way to Paris Comes upon a Traveling Companion, Arnold

THUS DEEP IN THOUGHT he met a traveler
To Paris bound, a bundle on his back.
To him Brunellus spoke and gave a kiss,
Asked who he was, where born, and whither bound.
He answered: "I'm from Sicily and on
My way to school in Paris; safe trip, I pray!"
To him Brunellus said: "We share three things—
Desire, a purpose, soil; let's share our trip."
Arnold agreed and since his comrade urged,
He let him carry all his bags and books.
Then joining hands in act of friendly trust,
They hastened on to Paris side by side.
And as they went along Brunellus told
Him who he was, his home, and of his trip,
Of dangers and misfortunes he had met,
How such and such a thing had been his lot,
How first he came to leave his native land,
The counsel Galienus gave to him,
Brunetta's fortunes and Bicornis' fate,
And how they faced both seasons of the year.
He told how at Salerno he received

Ten glassware vessels from a London cheat,
The way Fromundus set his dogs on him,
And how much loss that villain brought to him.
He then described the killing of the dogs,
The leap of Monk Fromundus, and God's wrath.
He added too the epitaph he wrote,
And also told why he was Paris-bound.
When his misfortunes he had thus made known,
His comrade Arnold then made this reply:

Arnold's Story about the Priest's Son and the Little Chick

HOW HUMAN fortunes shift and vacillate
'Tis not an easy task for me to say.
How from the slightest cause calamity
Can rise, the outcome clearly demonstrates.
Apulia saw a strange event take place
When William ruled, the present duke's grandsire.
A certain urban priest moved far from town
Where he began to live a country life.
And since he always tilled his lands with care,
He usually had very heavy yields.
He had a wife and children too, if one
May speak uncensured of a bishop's wife.
One son, Gundulfus, whom I've seen myself,
Since he was still a child, remained at home.
Since he was keeper of his father's grain
And watched the door, he often held a whip.
Now once perchance a hen, her brood close by,
Came to the crib to pilfer grains of corn.
Behind their mother came the hungry chicks

And in their boldness hurried through the door.
The boy then lifted up the whip he held
And started striking them and chased them out.
But since he grew excited, he transgressed
The bounds of reason with his shouts and blows.
When anger fails to see the better course,
It drives men on to many sinful deeds.
A careless throw oft brings a lasting loss,
And in a moment's time rich stakes are gone.
Now since the angered boy applied too hard
The whip, he chanced to break a chicken's leg.
The fractured leg, to chick and mother-hen
Alike, was long a cause of grave concern.
Both pain and pride enraged the chick so much
He wished he might get even with the boy.
As time went on the flesh grew back and healed;
The wound was covered over with new skin.
The broken bone brought long and grievous pain;
The wound had left its mark upon his heart.
Outside the wound was covered with a scar,
But inwardly raw wounds possessed his heart.
The leg an exile long was now restored,
The mind was exiled, brooding of the deed.
His step was straight, his hip was well and strong,
He noticed not the break, the limb was well.
His mind was maimed, his breast, though woundless, faint,
His heart repined, his mind was much aggrieved.
Delayed revenge allows the wounded heart
Some respite, but stands guard before the door.
No rest nor solace has the injured mind;
Revenge alone relieves the wounded heart.
The loon cares not for pond, nor wolf for lamb,
Nor dogs a hare, or captured beast escape,
Nor does a fish crave water, hawks the lark,

As much as wounded hearts desire revenge.
The cockerel too thus wished to get revenge
Against his foe Gundulfus, if he could.
The chick was now a rooster six years old,
His father dead and he now in his place.
Gundulfus too had now grown very tall
And was about to take his father's place.
Nought hindered now, nought failed him but one thing,
The need for him to hold the priesthood soon.
The bishop, yielding to his prayers and Saint
Rufinus' favors, bade him take his vows,
And set the day on which these holy rites
Be held and place where they should be performed.
The week preceding Christmas was the time,
The city Tarabella was the place.
Gundulfus' rise rejoiced his family,
And soon they planned a feast to celebrate.
His relatives, to have a family feast
Before the happy day, together came.
They all were welcomed, and the tables groaned
With food, and all indulged in meat and drink.
Much wine there was, in golden goblets served,
To cheer the guests and sanctify the place.
It was the night before Gundulfus planned
To start for town at dawn to be ordained.
At early cockcrow they desired to leave,
Because that place was rather far away.
The time was told the servants, and the hour
Was set for leaving at the break of dawn.
They meant to listen for the early song
The cock as herald of the morning sings.
It would be soon enough if then he rose,
Because the night was long, not bad the road.
The cock was wide awake and heard these words,

And in his heart rejoiced, but held his peace.
So pleased indeed was he by this that he
Could hardly keep his lips from shouting praise.
And what by silence he was eager to delay
He almost hastened by his shouting out.
A voice made still by grief cries out with joy;
Joy fosters lyrics; grief bids one be still.
Betwixt the twain he stayed, nor did he yield;
By neither overcome, his voice was still.
The night had almost all been spent in drink,
And overcome with wine they had retired.
The hours of night by frequent drinks beguiled
Sped swiftly by and almost unobserved.
But constant drinking could not keep the night
From binding them with its accustomed law.
Meanwhile the hour of cockcrow was at hand,
But still the crower uttered not a sound.
The crower's voice was still, the servants too.
Cups, night, and sleep forbade them stay awake.
A hen long wond'ring why the cock was still
And why he failed to carry out his task,
Drew near her mate and whispered in his ear
That both the time and hour had now passed by.
He thus replied: 'Don't bother me, hush up!
You'll always be a fool; away you fool!
Oh, woe to him that's wedded to a fool!
His bed will never be exempt from grief.'
Not less, but even more, she begged her spouse
To herald forth the stages of the night.
But he, opposing, tried to keep her still,
And now he begged and then he threatened her.
She swore, however, that unless he crowed
She'd sing and shake the rafters of the house.
And tired of waiting, from her throat she poured

A raucous cry, the best that she could do.
A certain one on hearing her replied:
'Stop, hen, I pray, for nothing do you gain.
Although a chicken cackle in the night,
No sooner will she make the sun arise.'
Meanwhile night swiftly passed, and all the house
Was deadly drunk, the guard himself asleep.
Deep sleep prevailed; Gundulfus also slept,
And in his sleep he had a pleasant dream.

The Dream of Gundulfus

"HE'D BEEN ORDAINED, he thought, and now was back
And dressed in priestly robes to say the Mass.
The cantor's place, he saw, the cock assumed
And led the choir clad in snow-white gown.
The clear-voiced cock, though 'gainst the will of all,
With this introit started off the Mass:
' 'Tis clear, O Lord, that all the things you've done
For us were by a righteous judgment done.'
And as the chalice filled with wine stood near,
He held it high, and all its contents drained.
And when he'd drunk the wine, and it was gone,
In his own way he finished all the rest.
But when the cantor was supposed to say
'Depart!' he said no word nor gave a sign.
On seeing this Gundulfus was amazed,
And crying out he asked: 'Is this the day?'
Though he desired to rise, his servants said:
'No, no! Stay where you are until it's time.
Not yet has crowed the cock who knows the hours
Of night and signifies each by his song.
Though we should wish to steal more time, the cock
Would indicate the hour in spite of us.

For he knows better than we do the hours
Of night, the number passed, the number left.
Let's leave to him the time for us to rise,
Let him who stays awake keep watch for you.
The night is long, the greater part remains;
Turn over then and sleep; there's lots of time.
A winter's night is like three days in length,
Nor can it in a moment fly away.
We'll be awakened by the faithful cock,
For he can't hold his peace, if he should try.'
At this the rooster whispered to the hen:
'I'll neither pledge nor promise you a thing.
Let well enough alone; the law provides
That he who sleeps shall sleep, who drinks shall drink.'
While he was saying this, the light of day
Burst through the windows and through every chink.[27]
The rays of Phoebus now reached all the earth;
The plowman now with oxen turned the soil.
As he arose Gundulfus smote his breast
And cried, 'Oh, how I wish that I were dead!'
In utmost haste (his pants left on the bed)
He ran to put the saddle on his horse.
The bridle and the saddle were misplaced;
The plowman had that morning taken them.
With but a strap he mounted nonetheless,
Nor came back to his father to be blessed.
Thrown off his mount as through the town he sped,
He fell to earth; his steed roamed far and wide.
He then pursued his way on foot, but came
Too late—the ordination was performed.
The bishop having finished everything

[27] "Clarum iam mane fenestras / Intrat, et ad rimas lux manifesta ruit." Cf. Persius 3.1f.: "iam clarum mane fenestras / intrat et angustas extendit lumine rimas."

Had left the altar; nothing more remained.
The lesson had been read, and he had said,
'But you'; the boy intoned the word, 'Amen!'
What could Gundulfus do in such a plight?
His only recourse was to go back home.
And so toward home he sadly turned in haste,
Bewildered, shamed, in tears, and sore distraught.
His father, mother, and his relatives
All wept and beat their breasts and wrung their hands.
The relatives who'd come returned distressed,
For they had seen their gladness turned to tears.
One blamed the guards, another those who drank,
One blamed the cock, another cursed the wine.
On hearing this the hen rose from her nest,
Went out and to the cock had this to say:

The Hen's Complaint

" 'OUR DEAR GUNDULFUS has returned from town
Frustrated, sad, and shedding many tears.
His parents, brothers, wretched sisters too,
Are bathed in tears; there's mourning everywhere.
Not even if Gundulfus had been hanged
Could they have mourned his death as much as this.
And for his plight they place the blame on you,
As if the fountainhead of all his woe.'

The Sentiments of the Cock

"TO HER THE COCK REPLIED: 'Not wrong, I think,
But right is it to render tit for tat.
Gundulfus struck me when a baby chick
And broke my leg; I still recall it well.
That he was first the source and cause of pain

To me, my broken leg remains as proof.
He was the first to laugh when I was hurt;
Now I in turn can scoff and laugh at him.
Thus fortune shifts, thus joy gives place to grief;
'Tis thus that lofty things are wont to fall.
Sweet medicine for pain, though long desired
And slow in coming, pleases when it comes.
It never comes too late to the diseased,
Provided that it somehow bring relief.
A vengeance for a wrong that comes with ease,
Preceded by no toil, more sweetly comes.
I've won a happy triumph without blows;
My mind wreaked vengeance, though my lips were still.
Let others fight; let silence be for me;
Let voice and trumpets ring; calm pleases me.
The sooner victory in war is won,
So much the greater is the victor's praise.
We need no arms when silenced voice wins out;
That one who feigns yet does not talk is wise.
Let tears, therefore, for blood drip from my foes;
Let grief and anger take the place of wounds.
Indeed a wounded heart will pain and burn
More deeply than will flesh when cut by sword.
A surface wound responds to medicine,
But wounded hearts will seldom ever heal.
An injured mind less quickly finds relief
Than wounded limbs respond to healing hands.'
No sooner had the rooster said these words
Than both the parents of Gundulfus died.
His father dead, Gundulfus soon was forced
From home, and gave up all his house besides.
Gundulfus, wretched, poor, a mendicant,
Forsaken, from the city went afar.
But many still recall this happening,

And fathers often tell it to their sons,
That mindful of the deed they'll so control
Their lives, that they need not, too late, repent."

How Brunellus Comes to Paris, and What He Does There

CONVERSING thus together as they walked,
They came to Paris and they found an inn.
Rest healed their weary limbs, while food and wines
Made good the losses of their meager fare.
Bones, skin, and muscles travel-worn and tired
Were freshened by warm baths and care and rest.
Brunellus got a haircut and was bled,[28]
And clad himself in more attractive dress.
When groomed and bathed he then came into town,
Went to the church and made his vows and prayers.
Then to the schools he went and asked himself
Which school would suit him better, this or that.
And since he thought the English quick of wit,
For many reasons he enrolled with them.
They have good manners, charming speech, good looks,
They have keen minds, and judgment that is sound.
They pour out gifts, and stinginess detest,
They serve large meals, and drink without restraint.
They hold gay parties, drink, and have their girls,
Three vices these in which they all partake.
Except for these you'll find no fault in them;

[28] "Brunellusque sibi minuit." Bloodletting was a very common practice in the Middle Ages. Vegetius *De Arte Veterinaria* 1.16.2 also uses the verb *minuere* without *sanguinem* with the meaning, "to let blood." Cf. Knowles, *op. cit.*, pp. 455–56.

Take these away, all other things will please.
Yet these should not be always criticized,
For there can be a time and place for these.
For two of them are largely free from pain,
And often lead to paths of happiness.
The third thing keeps the ferment[29] by which France
Is filled from being able to cause harm.
Hence wise was he in his desire to join
The Englishmen, to share their way of life.
Another reason too he had to wish
To join with them, for talk is but a dream.
If those who live together learn like ways,
Why not unite with them if possible?
If nature gives to them a better lot,
Why not derive from it some benefit?
In eager haste he therefore joined their school
To learn to speak with charm and by the rules.
But since his mind was dull, and stiff his neck,
He failed his courses; toil and pains were lost.
Brunellus had already spent much time,
He had completed almost seven years,
Yet absolutely nothing had he learned
Of what his master taught except "heehaw!"
What nature gave and what he brought with him,
That still he had, and none could take from him.
The masters, having labored long and hard,
O'ercome with weariness, at last gave up.
His back was often beaten by a club,
His sides were lashed, his hands endured the rod.
He always said "heehaw!" and nothing more
Could say, regardless of the kind of blow.

[29] "Fermentum." Cf. *Contra curiales*, p. 177: "Hoc est fermentum quod universam massam corrumpit," in which Nigellus is describing the evil influences of ambition.

One pulled his ear or jerked his crooked nose,
  Another knocked out teeth or pricked his hide.
They slashed him, burned him, freed him, tied him up,
  Sometimes they uttered threats, sometimes they coaxed.
Thus art and nature in him vied by turns;
  Art begged, while nature bade; art left, it stayed.
It's clear that those who have a background that
  Is weak can seldom, if at all, grow strong.
Brunellus learned as child "heehaw"; nought else
  Could he retain except what nature gave.
What nature gives remains, but what is learned
  By art takes flight like dust before the wind.
He lost his money, toiled in vain, and all
  That he had spent had likewise been for nought.
Hope too was gone of adding to his tail;
  He knew the claims the English made were false.

Confession of Brunellus for His Shortsightedness

BRUNELLUS then recalls his wasted youth,
  Its follies, and he thus reproves himself:
"Ah me for having lived! What madness drove
  Me here to Paris and into its schools?
Why need I study and do all this work?
  Could not Cremona have supplied my needs?
I scaled the Alps and left them far behind;
  A fool was I to go so far away.
Why did I leave my land and home, and come
  Across the Rhone to see new lands?
Why did I wish, with risks of death so great,
  To see the sons of France and Paris schools;
To find the English eager to refill
  The cups; the grasping French to offer threats?
Apulian came I here, a Gaul I leave,

Brunellus stays, however, just the same.
Not one thing have I learned; the little which
I knew before is all gone now, I'm sure.
If I could but retain and speak two words
Of French, I'd certainly be gratified.
And if perchance 'twere three or even four,
I'd think myself Jove's match or greater still.
I'd bring on Italy such dreadful fear
That tribute would be paid me by the king.
My trip to Paris then had not been vain,
If my own land should thus be ruled by me.
But this was not to be; far otherwise
Were spun my threads of fate from what I prayed.
My fates were surely very harsh to me;
They brought me many evils, nothing good.
And it's quite clear that there is not a lot
In all the world that's worse than this of mine.
I'm dull; my brain is harder than a rock,
My heart is more inflexible than steel.
I have a leaden heart and head and brain,
Indeed of metal heavier than lead.
My legs are iron, my sides like sheaths of iron,
In all my body there is not a vein.
Just like a brazen vessel is my skin,
Which may be beaten but receive no pain.
It's not for me to die from curse or blows;
With mallets could I scarce be killed, I think.
Why from the cursed womb did Mother bring
Me forth and sever not my throat with sword?[30]
If to her sorrow I had been still-born,

[30] "Cur mea me mater maledicto fudit ab alvo." Manitius, *Geschichte der lateinischen Literatur des Mittelalters,* III, 812, observes the close parallel of this verse with *Anthologia Latina* 786.1: "Cum mea me mater gravida gestaret in alvo."

How fortunate and blessed it had been!
Why did a hungry wolf not come to bear
Away the babe while still of tender age?
Why did my mother curse me in my youth
And often ask the wolf in haste to come?
Not only will I surely be his prey,
But greedy wolves, I think, will feed on me.
For swift fulfillment, if the saying's true,
Is always present for a mother's prayers.

The Dream of Brunellus

"I'M ALL THE MORE DISTURBED and filled with fear
Because last night it seems I had a dream.
I dreamed my parents offered prayer to God
On my behalf, and this is what they prayed:
'Be kind to us, O God, and from the jaws
Of wolves snatch our Brunellus; bring him home.
May he who is abroad come safely home,
Unharmed and unmolested by the wolves.
May neither lion, leopard, nor that worst
Of beasts, the rustic, stop him on the way.
In safety may he travel all the roads,
May he escape the scent of every beast.
May dogs be tongueless, ears of cats be closed,
And may the wolf have gout so he can't run.
For since we fear him more than other beasts,
From him protect Brunellus; bring him home.'
Now since dreams often mean the opposite,[31]
The fear that dwelt within my mind increased.

[31] That dreams are often to be interpreted in a manner opposite to their apparent meaning seems to have been a commonly held belief since antiquity. Compare Pliny *Epistulae* I.18: "Refert tamen eventura soleas an contraria somniare"; Lucan *Pharsalia* VII, 21,22: "sive per ambages solitas contraria visis / vaticinata quies magni tulit omina planctus."

What things we see as visions in the night
Need not occur as soon as morning comes.
But dreams, contrariwise, must be explained
The very opposite, I'd have you know.
If dreams are good, then trouble will result;
If bad, doubt not that great success will come.
'Tis thus my mother used to speak of dreams,
And she was very well informed and wise.
'Twas over this my mother often used
To quarrel with father, answering word for word.
For he with mother always disagreed,
Maintaining that her theories were all wrong.
Indeed he claimed true knowledge of the stars,
From childhood having learned the signs of heaven.

Brunellus Conceives the Idea of Becoming a Bishop

"BUT STILL, no matter what my parents thought,
I trembled and my mind was filled with awe.
But why my fears, since things need not be feared
Which can't in any other way be changed?
Whatever fate decrees, I have no doubt,
Will come to pass; no one can contravene.
If fate decides in human fortunes what
Shall be, then it must happen just that way.
No one is able to avoid his fate,
Or turn it from its course, nor can I mine.
And yet I know not what my fate shall be,
A happy one or hard, adverse or fair.
Perhaps I'm destined for the bishopric,
And will obtain that office in my land.
Indeed the world sees stranger things occur
Than my advancement to the bishop's chair.
And if the bishop's vestments I should take,

Where could the miter sit upon my head?
With ears erect, in style pontifical,
For miter there will be no place at all.
The miter holds not all the bishop's power,
Though it's the sacred emblem of his rank.
The miter then does not concern them most,
But rather it's the power that with it goes.
The miter shall not gild this head of mine
Without the bishop's powers, so help me God!
The miter nor its horns shall crown this head,[32]
Unless there's present what accompanies it.
When all things else are lacking which belong,
What joy is there to hold the empty form?
Full bishop I shall be, for I don't want
His honors as a mule, but as a horse.
Be ever far from me the abbot's ring,
His pompous cap, and honor without God!
The abbot's sterile miters with no part
In chrismal rites I hate—away with them!
The mule has genitalia, yet he
Is always sterile and can't reproduce.
So those who have no office, but the name,
Have nothing but insignia to bear.
Forbid that I should rise and take such horns
As various ones have for themselves assumed.
I want to be a bishop with full rights,
And not to be a bishop just in name.
They have the name without the office who
Take horns, with nothing else that they can show.
Forbid that I do this or that I take

[32] "Neque cornua sumam": Manitius, *op. cit.*, compares Ovid *Ars amatoria* 1.239: "tum pauper cornua sumit"; but the similarity seems to be one of words only. *Cornua* in Ovid is used figuratively for "strength" or "courage." Here the word is undoubtedly used by metonymy for miter which was adorned with horn-shaped corners.

The horns reduced in any kind of way.
I'd rather be content with both my ears
Than have two horns like that appear on me.
In what things they excel, for they fall short,
The mule and abbot are in like esteem.
The fillet means the same on abbot's head
As signs of masculinity on mules.
Those who from being bishops are deprived
The flattened miter does not abbots make.
They have an idle name, bear specious signs
Wherein full meaning of the name is lost.
Brunellus will not thus assume high rank,
He'll not let useless horns on him be placed.
No servile gift nor hasty promises
Nor hand unclean shall blot his name.
He'll not be forced, nor through foul simony
Make bargains, neither will he give rewards.
He'll make no promises, for never should
A bishop's mind be seared, but always sound.
No begging, bribing, cringing to the strong,
Will bring Brunellus to the bishop's seat.
Sincere and sound of mind shall I proceed,
Believe me, if I never reach that goal.
I'll take the proper course and when I'm called
To service go that none may censure me;
Lest after such an honor's been bestowed,
Someone might say, to cast reproach on me:
''Tis thus you've risen, thus advanced, and thus
Through me have come a shepherd to your fold.
To rise to bishop you have pledged me things;
Now keep your word, why act like this or that?'
The life of bishop is a book which each
Should read aright and follow as a guide.
The clergy and the faithful ought to keep

The bishop's life a pattern for themselves.
No fault nor baseness should in him be found,
But he should be a model to his flock.
Let him not lightly bind or loose,
Let every power he has be highly prized.
Let him be honest, frank, and soft of voice,
Intent to honor not to burden men.
He who should freely give should never take
By any means the things another has.
For if the eye of mind be dimmed, then all
The body will be darkened, and all life.
If all the virtue of the world should on
One bishop rest, 'twould hardly be enough.
Let neither hate nor love make dull his mind,
Nor let his inner eye be dimmed by mist.
Not flesh nor blood, but Holy Spirit should
Dictate what first the bishop ought to do.
Let him not act through love of flesh, or wrath,
For wrath repels, while love restrains and holds.
Let him be helpful to the lame, the blind,
The poor, the fallen, and to those condemned.
Let widows' griefs and sorrows touch his heart,
And let him deem the poor man's tears his own.
Let him take care of orphans and regard
As lost what is not given to the poor.
Let him not spurn the poor and praise the rich,
For man beholds the face, but God the heart.
Let him not seek the people's wild acclaim,
Let him strive not to please the world but God.
For what else but a fleeting gust of wind
Is glory, wide acclaim, or brief esteem?
What joy is there in winning fame of men
If you deprive the poor but aid the rich?
To fill the rich man's stomach with the poor

Man's tears, I find no cause for praise in that.
It's better to cast fame aside when it
Works mischief than to win it by foul means.
The many who revere men's praises for
An hour deserve to be dishonored long.
He who to win excessive praise of men
Aspires too high is acting like a fool.

Brunellus' Foolish Boasting

"NOW WHEN I'M RAISED to bishop in my town,
I'll have no equal in the whole wide world.
The people from the city all will come,
And bowing low will say, 'O Bishop, hail!'
What can my mother say, I ask, when she
Sees clergymen and laymen blessing me?
With joy she'll bless the year, the day, and hour,
In which she bore in blessedness her child.
How honored too and glad will father be
When lord and father of a bishop called?

The Return of Brunellus to Himself

"THE RUSTIC WILL APPEAR perhaps to see
My parents, now retired, and shake his head.
He'll speak obscurely, for a shrewder one
Than he cannot be found this side the sea:
'Great things I saw when young, and greater still
Shall see, if I am spared and life remains.
The age belongs now to Brunellus, once
It was another's; changes thus take place.
The life of this great bishop now is in
Esteem, which was before quite different.
His mother who now spurns the common things

Will bend beneath her usual load, I think.
His father too will bear again the pack,
As bishop's father cease to be addressed.
The many things we've often seen and see
Let's still endure and not be moved thereby.'

Account of Brunellus on the Ingratitude of the Mayor of the City

"I HAVE GOOD REASON too to fear, since I
May rightly fear—a little thing can harm.
The mayor of the town, unless perhaps
He's forced, won't welcome me—he has good cause.
The truth is that he stole a sack of meal
From my own father; 'tis a deed well known.
The news spread far and wide through all the town,
All bear me witness, even God himself.
There came together to my father's house
Three thieves of whom this very man was third.
With ill intent they came by night and planned
To carry much away, if it were possible.
The first one took a horse, a sheep the second,
And he who now is mayor took the meal.
When they had left the house and started off,
And tried to make a sudden get-away,
The dogs, I know not if it were by chance
Or sound of feet, were roused and blocked their path.
In haste my father came upon them there
And bound them hand and foot with heavy chains.
When morning came the citizens were called,
And those engaged in practice of the law.
The thieves, released from prison and unchained,
Stood there, each holding his disgraceful loot.
Fear made them pale, respect for parents sealed

Their lips and caused their memories to fail.
Long urged, at last they started to confess,
Although their guilt was fully known before.
Confession made, the citizens decreed
The culprits, just as they deserved, be hanged.
All through the night the servants readied ropes
And crosses which at dawn each must ascend.
Compassion then which had increased in me
From birth caused me to pity these poor souls,
Though I recalled that old familiar saw
Applied to situations such as this:
'Among a thousand men that man is worst
Who shall decide to steal you from the cross.'
But still, although I shrank from this, at last
Compassion overruled and mastered me.
In secret then I took my father's keys
And did a valiant and still vivid deed.
I loosed their chains and led the thieves by night
From prison cell, alone and unperceived.
For had I been detected in such crime,
Forthwith due penalty I would have paid.
And so, because I intervened, escape
They had alike from blows, abuse, and death.
And that no one by chance find out, I bore
Them on my shoulders over streams and land.
The three, that none of them be left behind,
I carried far away upon my back.
And though I've learned to carry heavy loads,
I don't recall a burden such as that.
Each one alone was quite a heavy load;
How much the more did all together weigh?
If I shall live five hundred years, I'll be
The worse for bearing that enormous load.
No day will pass but I shall call to mind

That weight I carried, for it bore me down.
But I preferred to save the wretches by
My toil than let them perish on the cross.
And as they bade farewell, they fell upon
The ground, and bathed in tears could barely say:
'O Lord Brunellus, we your servants are,
We're yours forever and in every way.
We're yours to sell, to call, or go
Wherever it may please you that we go.
Us wretches three you've freed and given life,
A gift not ours but yours will ever be.
Your servant each forever will remain,
Nor will he ever cease to be your own.
'Tis right that he should be your servant true
Whom you have saved from prison, noose, and cross.
Shall he whom you as lord and master deigned
To carry on your shoulders not be yours?
Forbid that he should not be yours, for he
Could never have his freedom 'gainst your will.
And if our life continues we'll reward
Your kindness with a fair and due return.
A rich return we'll give you for your help
If God will grant us power to pay you back.
May God grant us the opportunity,
Nor let us die until you've been repaid.'
I well remember this, for many times
They knelt and said, 'Behold we three are yours!'
To hide my crime from father and the town,
As soon as they escaped, I fetched their chains.
My sin, indeed my greatest sin, was this:
I claimed the chains were broken by the power
Of God, that Leonard, saint of God, had come,
Broke loose the chains and carried them away.
So now I fear lest, as the people say,

The mayor may requite me for those deeds.
He may attempt to render ill for good,
And wish me not his peer who helped him once.
In this I'd not lose hope nor rightly fear,
Since I am conscious of the crime he did,
If in that rogue by nature there could be
Implanted just one drop of noble blood.
For this is proof of noble blood, to wish
To make a due return for benefits.
To strive with nature and its canine ways,
How difficult it is, no one can say.
To wipe away the marks of servile rank
Cannot be done by doctor nor by drugs.
The product always shows and cannot hide
What nature really is and whence it came."

Departure of Brunellus from the City of Paris

BRUNELLUS having meditated thus
From Paris planned to go next day at dawn.
So having bade his weeping friends farewell,
He made his leave and hurried on his way.
When he had reached a mountain top and saw
The city left behind, confused he said:
"Thy blessing, Holy Mary, God, and cross
Of Christ! What might that be in yonder vale?
Methinks it's Rome, with mighty walls begirt.
What else but Rome, this city of such towers?
And what can that be but the mount of Jove?
O God and Holy Mary! Rome so near?
The city where I studied now has slipped

From mind and lips; what am I anyway?
If I go home now, and my parents ask
Where I have studied, how shall I explain?
Perhaps they'll say I'm worthless and have failed
To learn a thing, but wasted all their means.
So I'll go back and learn the name I've lost,
Lest I be called a fool when I get home.
It will be something just to name the towns
And to pretend I've studied in their schools.
But if I cannot even give the names,
Not one thing which I say will they believe."

On Brunellus' Silence

As he thus spoke and wished now to return,
A rustic met him and addressed him thus:
"Brunellus, tell me why you're Paris-bound.
Do you go there to study or to teach?"
No sooner had Brunellus heard the town
He turned and held upon his lips the word.
And lest he lose again the name, or by
His prattle have it happen as before,
He made resolve to utter not a word
Save "Paris" only for the next fortnight.
No matter what occurred or whom he met,
He'd make no answer, but appear a mute.

How Brunellus Breaks His Silence

It happened meanwhile that when he had spent
By now a dozen days in this resolve,
A stranger from a town nearby and bound
For Rome became his comrade in the Alps.
Approaching him he said: "O reverend sir,

The Lord be with you, may your day be good!
Success to you today, an evening fair,
A pleasant lodging, and a restful night!"
Brunellus who till now had held his peace
Made no reply to him nor said a word;
Yet humbly bowed and made acknowledgment
With nods and signs, but not by word of mouth.
That night they came together to an inn,
Where they retired to rest their weary limbs.
And as the stranger uttered time and time
Again the Paternoster as he prayed,
Awake, Brunellus said: "The stranger's speech
Has robbed me of the only word I knew.
A corresponding syllable so oft
Repeated robbed me of my city's name.
The Paternoster said so many times
Has caused too much my thoughts to turn to him.
The stranger by a word so much alike
Distracted me and took away my words.
It's wiser then to sever social ties
Than to an evil partner be allied.
My mind, a truthful seer of coming doom,
Had prophesied that this would be my lot.
Exchanging my own word for that not mine,
I lose my hard-earned gains of seven years.
The little word which perished is like that
The stranger used, in its first syllable.
The end was not the same, and if I could
Remember it, I'd not yield place to Jove.
If I were rich and had a thousand pounds,[33]

[33] The Latin word is *solidus*. Strictly speaking the *solidus* was a coin perhaps equivalent in the Middle Ages to a shilling, but the word is used in a very indefinite sense in this context, and may, I think, be translated here by the word "pound" without any violation to the sense.

I'd gladly give it all for such a thing.
The first part of the name is nearly gone,
And what remains has ceased to be my own.

Consolation of Brunellus

"IT'S BETTER to keep something than lose all;
The only part that's left, the first, I'll keep.
The wise have taught to use a part for all;
If it remain, one part's enough for me.
A thing's not whole, when part is cut away,
Nor is it wholly gone, when part remains.
It's better from the whole to keep one part,
Whatever it be like, than lose it all.
Perhaps I couldn't keep the larger part,
For it is lost, but this one part I'll hold.
If from the seven arts one syllable
Is left, I deem it not too small a thing.
And as instruction often makes men err,
It also causes them to be puffed up.
Lest knowledge of a letter bring me ruin,
One syllable's enough for me, no more.
Sometimes superior training breeds disgust,
And learning makes men's character austere.
More often nature serves you better in
A few things, and more strongly shows her power.
Those satisfied with just a little, gain
Thereby, while constant care disturbs the mind.
To be informed of many things but nothing well
No honor but a burden surely is.
A little many know of many things,
Yet on the whole appear quite ignorant.
To know is hard, and also to retain;
To learn is hard, and hard it is to teach.

And I am all the more disturbed because
  No discipline can teach there is no death.
Does fear of death not reign through all the world,
  And does it not exist in every land?
With even balance of the scales it comes
  To all alike, relentless but most just.
I can't refrain from fearing it will come
  Quite suddenly and say to me, 'Away!'

Brunellus' Remorse

"I'VE THUS RESOLVED to take religious vows,
  That I may win salvation for my soul.
And to redeem my youthful years, it's right
  That I now old should seek to save my life.
There's left a little time of life, and lest
  I waste it all, I wish to end it well.
'Tis better late than never to repent;
  'Tis right that one ashamed of sins repent.
And if the morn and all the day be dark,
  Clear eventide may crown the cloudy day.
What heedless youth has scattered recklessly,
  A waster of itself while in its flower,
Let ripe old age reclaiming youth's mistakes
  Use its own flowers and gather in its fruit.
Hope rested in the flower, but then it drooped;
  Hope passed away when fruit there failed to be.
The promise of the flower is gone, but may
  The one last hope of fruit still left bring good.
Than death, so sure a thing, there's nothing more
  Concealed; it's hid from all, extends to all.
In human fortunes death comes to us all;
  Each one of us takes turns at ownership.
If anything belongs to one or all,

There's nothing more for one or all than death.
What do I here except to serve the world?
Tomorrow death will tell me, 'Rise and come!'
There's nothing then for me to do but take
Religious vows; away with all delay, I pray!

Brunellus' Feelings toward Various Existing Religious Orders

BUT SINCE religious groups are numerous,
I do not know which one I should select.
If to the church I go with the red cross,[34]
They'll send me overseas to pay my vows.
A slave I'll be and serve in distant lands.
I'll serve perhaps around the walls of Tyre.
But I'll not go on foot, but on a horse,
Oats-fed and fat and soft and light of foot.
And let him watch his step and stroll along
With gentle gait according to our rule.
The rule forbids me mount a trotting horse,
Nor do I wish the rule relaxed for me.
I'll go as knight, that I may wear white cloaks,[35]
And then I'll go to war, but not return.
The Sultan will possess my leather reins,
And savage beast will feed upon my flesh.
The heathen's sword will tear my vitals out,
And he will flay me and inter the rest.

[34] The red cross, it is believed, was originally the insigne of crusaders. See *Catholic Encyclopedia*, IV, 537.
[35] The Knights Templar wore white habits.

Concerning the Hospitalers[36]

"AGAIN, if I shall be a white-crossed knight,
I'll be dispatched to Lebanon for wood.
In tears I'll go, and slashed with stinging whips.
My stomach will be empty and dried up.[37]
Though much befall me, I will have no rights
Except to magnify the house with lies.
And if I break the rules just once or twice,
They'll say, 'Get out!' and take away my cross.

Concerning the Black Monks[38]

"IF I SHOULD WISH at Cluny to become
A monk, they'll feed me eggs, black beans, and salt.
They'll force me to arise at midnight, though
I wish to keep my body warm in bed.
And what I'd rather not attempt to do,
They'll want me to rise up and sing a tune.
My voice is very harsh, although it's loud,
And blockage of the chest provokes a cough.
If I won't sing, they'll make me bear the lamp,
And in the morning give me my just dues.
Rich foods and meat they often may consume,

[36] Wright, *op. cit.,* p. 82n, says that MS B has the rubric, "Ordo sancti Johannis Jerusalem," and that MS P adds "alba cruce signatis" (marked by a white cross). This order was indeed known by the two titles, the Hospitalers, and the Order of Saint John of Jerusalem.

[37] Wright's reading *profinellus* has, according to Mozley, *Archivum,* IX (1935), 95, seventeen variants. Mozley accepts the reading *prope nullus,* which gives approximately the meaning I have given. Otto Skutsch, *Archivum,* XI (1937), 30, emends to *cophinellus,* "hay-basket," which would mean that both the stomach of Brunellus and his hay-basket will be empty. This is an interesting possibility.

[38] Wright, *op. cit.,* p. 83n, cites the following rubric from MS B: "Ordo Cluniaci" (the Cluniac Order).

And these on Fridays they may often have.[39]
They wear skin shirts and many things conceal
Which they keep hidden from their fellow-monks.
Their lords, the abbots, too are unaware,
Though very eager to enforce the rules.
And if the rule of silence be transgressed,
They'll reprimand, and use the whip besides.
And if I strike in heat of wrath or wine,
They'll charge me heavily and lock me up.
For only those within the cloister, tried
By much long service, do they let go out.
No love of abbot there nor family
Can have effect against the cloister rules.
Expenses of the order or the life
The Black Monks are not able to sustain.
They do not buy or lease their property,
But everything is given to them free.
What cells they think disgraceful and in ruins
Supply to many large commodities.
If from the members there arise a cloud,
The abbot will resolve it honestly.
And once he's passed a sentence on the crime,
No one is after that allowed to speak.

Concerning the White Monks[40]

"IF I BECOME a White Monk, they will serve
The usual two hard foods, but thoroughly cooked.
They hate what I unceasingly dislike,
A rustic working in his fields nearby.

[39] "In feria sexta saepe licebit eis." MS B, according to Wright, has the reading "In feria quarta non licet illud ibi" (On the fourth day of the week that is not permitted there). Again, p. 119 (Wright p. 95), MS B has "feria sexta." I have followed Wright's text, as does Knowles, *op. cit.*, p. 677.

[40] The White Monks were Cistercians.

Desirous of more land, their neighbors' bane,
They never want to have fixed boundaries.
Their sheep and cattle give them all the milk
And wool they need, but they're not satisfied.
Content with little yet they strive for more,
And though they have all things, they always want.
They'll not allow me any leisure there,
But feed me well in my novitiate.
They all are given work, that there be found
No indolent or idle person there.
They seldom keep the Sabbath; food is poor;
The lash is common there, and life is hard.
They eat no meat unless the abbot or
The one in charge in pity gives consent.
Since they are kept from flesh of quadrupeds,
And may not eat it since it's bad for them,
They like the flesh of bipeds such as run
Or fly, not that it's better, but more rare.
And when they do eat meat, the neighborhood
Will be no witness of the smoke, nor be aware.
When meat has been consumed, you'll see no trace;
The bones lie buried, lest they cry, 'Look here!'
Three tunics and two cowls are issued them;
They all receive a small-sized scapular.
They have no drawers to bother them by night
When in their beds—no fear have they of that!
Their lower parts which breeches never touch
By day or night will always be exposed.
So what am I to do if winds arise
From out the south and quickly bare my rear.[41]

[41] Curtius, *op. cit.*, pp. 433–34, says that nothing was so amusing to the medieval mind as involuntary nakedness, and that this is a very common literary motif. It belongs, he observes, to primitive folklore, but also has literary authority, and is found in the Old Testament (*Genesis* 9.22; *Nahum* 3.5; *Jeremiah* 13.26).

How might one ever face so great a shame,
  And to the cloister then return again?
For if perchance my nakedness be seen,
  I'll never be a White Monk anymore.
And they can do without me, and they should,
  So that the thing I bear shall be unseen.
Offenses to the weak should be removed;
  Mine then should surely be, for I am weak.
There's greater need to hide more shameful parts,
  And to less worthy parts more honor give.
A brother may conceal his nakedness
  Or not, but he who covers it is wise.
What you wish not be done to you, don't do;
  What you wish done for you, you also do.[42]
A brother is so-called not just by name,
  But deeds; the name without the deeds is vain.
We know that many whom we wrongly call
  Our brothers are like enemies to us.
Since Paul among false brethren had endured
  So many risks, he names them in his works.
Real dangers from unfaithful brethren rise,
  And from false brethren every evil comes.
Let each against his brother be on guard,
  Nor let him risk his hope and faith in him.
Fraternal trust is rare, for though each one
  Be true, he proves untrue by his distrust.
The carnal brother is deceived as well
  As spiritual; his brother each deceives.
If you should find a brother who is true,
  Let him become to you your other half.[43]

[42] "Tu tibi quod non vis fieri ne feceris ulli, / Quod cupis ut faciat quis tibi, fac et ei." Cf. Columbanus, *Distichs*, 24.25; Baehrens *Poetae latini minores* 3, 241: "Quod tibi vis fieri, hoc alii praestare memento. / Quod tibi non optes, alii ne feceris illi."

[43] See Note 6, above.

Indeed so many are the tricky, false,
  And vile, the world can't longer number them.
They dwell in cloister, as in Paradise
  Dwells Satan, where he surely has no right.
This frightens me as to religion's fort;
  I go and heed not where I rush or whence.
Perhaps my fears are vain, because the hearts
  Of men are often moved by idle fears.

Concerning the Order of Grandmont

"If I shall take the Grandimontine vows
  And wear their dress, I fear a sterner life.
For they have nought, nor care to have
  A thing; with nought they're ever satisfied.
They live secluded and no silence keep;
  They know not signs; their tongues are unrestrained.
No lands or fields nor pastures wide they wish,
  Nor is their annual yield of wool ten sacks.
They know not how to grade their wool in threes;
  One sack contains the breast, the neck, and sides.
They do not tan the pelts when stripped of wool,
  Nor send their swine to fatten in the woods.
They do not sail the sea to cast their nets
  For fish or to ensnare them with their hooks.
What foods they eat, how much they eat, a rich
  Or skimpy diet, that I would not know.
If what they have like manna ever falls,
  What benefit would come from wide estates?
If they like men and not like angels live,
  Why do they thus refrain from men's pursuits?
And if they're men, by nature mortal men,
  Why should they not abide by human laws?
What's done in secrecy can seldom be

Without suspicion, or unmarked by men.
If they essay to tame their stubborn flesh,
There is a better and more fitting way.
But I blame not nor anyone condemn,
For servant with his master stands or falls.
With many legal suits they vex the town,
For many days a simple case drags on.
They run up costs, though nothing do they own;
A case long buried lasts for fifteen years.
Paired off in two's, they struggle hard and long,
And they are unconcerned by needless costs.
For while, contrariwise, the layman rules,
The priest his sacred vows of office keeps.
On this account the mountain moved to Rome,
Yet there the case did not attain its goal.
They've wasted much, but fertile is the hill,
And rich, and with its milk fills all their needs.
Though no one plows or sows or reaps, it grants
In wondrous ways whate'er their hearts desire.
Why then is man, an angel inwardly,
Repulsed so often in the market-place?

Concerning the Carthusian Order

"SUPPOSE I CHANGE and join the Charterhouse?
I'll have some skins and several tunics there.
A cell I'll have in which to live alone,
No roommate nor a servant shall I have.
I'll sing alone, alone shall eat my meals,
And without lamp I'll find my bed alone.
I'll always have a three-room cell alone,
In which there'll never step a foot but mine.
There'll always be at hand inside my cell
Wood, beans, and water, for preparing meals.

The prior oft will come to visit me,
 And I shall have hot bread in ample store.
A certain share of sheep and herds is given them,
 And woods and fertile land on which to live.
Their underwear is made from wool of goats,
 And by its roughness castigates the flesh.
Thrice weekly bread and water is their food;
 They eat no flesh except on holidays.
And once or twice a month they may attend
 A Mass, if they have time and so desire.
Content with what they have they do not wish
 To burden rich or poor at any time.
They must not multiply their herds or lands;
 The cloister rules determine what they have.
They all are ever kept from eating flesh,
 Except for one whom leprosy consumes.
They settle not their quarrels in market-place,
 Nor do they render vain the people's 'Hail!'
They welcome guests and share with them their food,
 They give them what they have with cheerful heart.

Concerning the Black Canons

"BLACK CANONS likewise lead a worthy life,
 And suitable to God, as men believe.
They dress in soft but never rich attire,
 At meals they dine on flesh, but without wine.
What wrong has flesh done more than lentil, pea,
 Or bean, that it not be, like these, a food?
The guilt lies not in flesh which we condemn,
 But in wrong diet and in gluttony.
The wine is blamed; the fault must be with him
 Who drinks, when head and body ache next day.
Lest hymns and constant reading weary them,

They moderate their reading and their hymns.
They do not strive to see if they can burst
The lofty roofs with voices uncontrolled.
A linen wrap, pure white and finely spun,
And clean, these men so clean themselves enjoy.
Their dress is most distinguished and does not
Because of roughness scratch the flesh or skin.
The hem, however, is of black design,
To set apart the color of the rest.
The black, to please the eye, tones down the white;
And white brings joy in time of darksome fear.
Thus heaven's vault with double splendor shines,
And thus the pard has fur that's black and white.
White plumage gives black crows a brilliant coat,[44]
And likewise magpies are in black arrayed.
If flesh but not the heart two colors clothe,
The mind will be unstained, though stained the cloth.
Of what importance colors in one's clothes,
If mind unstained controls the inner self?

Concerning the Premonstratensian Canons

"PREMONSTRATENSIANS who come arrayed
In white are liked for their simplicity.
They have one color which is white as snow;
A simple sheepskin always covers them.
Since all are poor, till now they have agreed
To this, to wear coarse dress instead of soft.
The wool which stretches over naked necks
They set aside and use it for themselves.
Lest they indulge the flesh, they castigate
Their bodies with a wether's crusted pelt.
Yet as the order stresses, it is meet

[44] The European species of crow has some white plumage.

In place of linen to use downy wool.
The order is austere, for it forbids
At any time the use of flesh at meals.
And yet by dispensation it provides
Sound policy in this, nor is severe.
It grants them fat which is close kin to flesh,
But so that flesh be never evident.
That there be war with flesh and peace with fats
They've thus decreed, for they are fond of peace.
They thus desire to keep respect for flesh,
Lest unrespected be the law 'gainst flesh.
If flesh be fully banned, respect for flesh
Could well complain that it had been outraged.
If with no limitations it were served,
It would be constant cause and source of ill.
Hence it's not fully disallowed nor banned,
But roving exile in its native land.

Concerning the Secular Canons

"THEN THERE ARE CANONS known as seculars
Who've earned that title from their deeds.
Whatever suits their fancy is the law,
And under such a law they all abide.
These claim that nothing should be shunned or scorned
Which can be made obedient to the flesh.
And this especially they teach should be
Observed by all through all posterity:
That as the ancient law provides, no one
Should lack one girl, but each might have his two.
These love the world and grasp its fading flower,
And keep it watered lest it droop and die.
These do whatever wanton flesh commands,

That for their sins the way may smoothly run.
With them to guide, the entire world is led
Astray; they go ahead and speed its fall.
These hold the world, the world does not hold them,
Through them it downward goes to utter ruin.
These quick to undermine the roots of faith
By sinful acts make void the things they teach.
They cause the strength and power and rule of priests
To shake, they wreck the standing of the church.
Respect for clergy fades and dies through them;
The glory of religion comes to nought.
Through their designs the kings plan sinful acts,
And they are lax in things they should hold fast.
These are the king's and bishop's evil sides,
Their erring foot, false tongue, and thieving hand,
False heart, feigned love, and name that's dead,
A wrath concealed, real action, and false peace;
A bag without a bottom, nameless thief,
Unbalanced scales, true steelyard of deceit,
A godless law, a canon void of Christ,
Prime cause of evil, and a page of guile.
Impartial justice have they put to shame,
And taught their tongues to utter falsities.
While these desire to hold the slipping world,
They also slip and fall along with it.
Their life is toil, like dung their slippery path,
Unsure their end, yet pain their sure reward.

Concerning the Good Canons

"WHEN CERTAIN ONES of these mark well this truth,
They strive to keep themselves from such pursuits.
Earth's verdure and its flower they cast aside,

And count all things the world possesses nought.
Though in the furnace with its raging flames
They stand and bravely show their ardent zeal,
Yet they're not burned, because no fire without
Can injure those with spirits calm within.
The world embroils them, and as gold is tried
In fiery furnace, so does God try them.
To be engulfed in flames, yet be not burned,
Is not an act of nature but of God.
To carry on long struggles with one's flesh
Is difficult and greatly to be feared.
For mortal flesh like wax is swift to yield,[45]
And turn where old and wicked nature pulls.

Concerning the Moniales

"An order too there is of women who
Wear veils, and whom we call the holy nuns.
This group is formed of widows and young girls,
And to most people they are quite well known.
Their hours in usual manner they observe,
And carry out their tasks both day and night.
So loudly and so sweetly do they sing
That you might think you hear the sirens sing.
They're serpent-bodied, siren-voiced, with breasts
Of dragons, Paris' heart, Susanna's charms.
But still they have that which deceives all things,
A ceaseless flow of tears before their God.
With these they pray to God and win his grace,
With these they cleanse their hearts, whate'er they do.
They all may let their hair grow to their ears,

[45] "Nam caro mortalis levis est, et cerea flecti." Cf. Horace *Ars poetica* 163: "cereus in vitium flecti."

But only there, because the rule forbids.
They take the pelts of lambs before they're shorn
To clothe themselves; their habits are in black.
Beneath black veils they all conceal their heads,
Beneath black skirts they hide their lovely legs.
No girdles do they wear, nor underwear
In former times—if now, I do not know.
They never quarrel except when circumstance
Compels, nor strike unless they have good cause.
Some bear no children, others do; and yet
They hide it all beneath the virgin's name.
One who is honored by the shepherd's staff,
Brings forth indeed a more prolific brood.
Scarce one of them is found who can't conceive
Till age denies them this ability.

Concerning the Order of Simplingham

"THERE STILL IS ONE more order very new,
And, since it's good, it needs to be described.
In England it arose, and from the place
Where it was founded has received its name.
It's Simplingham, and from *simplicitas*
Or by antiphrasis received its name.
This house has canons, likewise laity,
And twice as many sisters too are there.
The canons only say the Mass; the rest
The sisters do, performing each her task.
A wall their bodies, not their voices, bars;
They sing in unison, though kept apart.

Brunellus' New Order Formed from the Other Orders

AS I in silence to myself thus muse,
  I do not know which life I should adopt.
And so I think it well, and wiser too,
  To organize a new society.
From all the orders I shall therefore take
  Whatever I find most appropriate.
And may this order take its name from me,
  And from it may my name forever live.
Let's take the Templar's smoothly stepping steeds,
  In order that my life go smooth for me.
Wherever I may go let me deceive;
  This privilege grant me from the other monks.
That I may be permitted to eat fats
  On Fridays,[46] this may Cluny grant to me.
From other brothers it is well and good
  That I be free at night from wearing drawers.
The Grandimontine monks I much approve
  For talking much, which thing I wish to keep.
Carthusians we ought to imitate
  Because they think one Mass each month enough.
Let's follow Black Monks in respect to meat,
  That in my order there be no deceit.
Let's imitate Premonstratensians
  With tunics soft and fully plaited robes.
I'm pleased to borrow from the group that's left
  A woman as my mate in lasting bond.
This order was the first, begun in Paradise,

[46] See Note 39, above.

Established by the Lord, and by him blessed.
We would that this be kept forevermore
To which belonged my mother and my sire,
In which my entire race has always been,
And if it fail, the human race will end.
From these devoted women I shall learn
To keep a girdle far removed from me.
Wide girdles are not good for me, nor are
Tight girdles suited to my bulging waist.
There still is one more thing respecting them
We'd like brought in our house when opportune.
I know not what I'll take from Simplingham,
Because its newness makes me hesitate.
But for the present this I shan't allow,
Indeed it seems most needful for the monks,
That never save in utmost secrecy
A sister with her brother may reside.
And there are things besides which if we can't
Recall right now, there'll be a later time.
So all that's left is confirmation by
The Pope, which he will grant me willingly.
For those who seek just things are never wont
To be rejected by their lord, the Pope.
It's proper therefore first to journey there
And humbly beg the brothers and the Pope."

How Brunellus Meets Galienus

AS HE was speaking, Galienus met
Him on his way from town, drew near, and said:
"You're not Brunellus, are you?" He replied:

"Ah me! I surely am, dear master, hail!
And from Cremona too, and in distress.
If you will try, you should remember me.
You see me old and crushed by ten year's toil,
Although I once was young and strong and brave.
How many ills have I now borne! And yet
The cares of school have wholly ruined me.
The rustic and the school are instruments
Of pain and anguish both to heart and flesh.
The rustic pierces, strikes, and beats the flesh;
The school consumes the vitals, heart, and lungs.
I'd rather carry rocks and stones in mills
Than study constantly inside the schools.
How sad my lot, and yet I cannot change
That which the fates have portioned out for me.
My soul is tossed about by doubts and griefs,
And from my grief I am bereft of mind.
Indeed so many are the sins of earth,
So many crimes of kings, of bishops too,
So many murders in the church and out,
Vile deeds so great they may not be revealed!
While I in tears thus note and see earth's woes,
What is, has been, and is to be its state,
As wax exposed to heat soon liquefies,
And snow and ice are melted by the sun,
In such a manner I break down and weep
And like a fountain pour forth floods of tears.
The chief and greatest cause for these stems from
The court at Rome as it now operates.
It was the queen and diadem of kings,
The city's fame and honor, and the world's.
It was a ray of sun, bright star of night,
A blow to sin, the scourge of evil men,

The sword of justice, mercy's flood of oil,
A lavish hand in need, with blessings filled;
Fresh flower that fadeth not in course of time,
Unfailing spring, and richly flowing balm,
Religion's crown, concord of peace, of high
Repute; the torch of faith, and faithful guide;
Our country's banner, yet it should have been
By all made head, just as it was the source.
But having turned around, without regard
To faith and name, it made its tail its head.
If *caput* you should claim from *capio*,
Then it is *caput* since it captures all.[47]
By conjugating thus, 'I take, you take,'
It stretched its nets, extremely wide, to take.
When I evolve the future from the past,
I disregard the present and the whence.
Once lavish and not grudging of its blood,
It often shed it for its fellow men.
But now, contrariwise, it thirsts for blood,
It sheds another's blood and drinks its own.
Thus she who should have saved has brought about
Her people's death and freely spilt their blood.
She shifted urns and poured from this to that,
With bile instead of honey rimmed the cups.
Yet she will always have insatiate thirst,
And though she ever drink, she'll be athirst.
Although a mighty sea of money flow
Into her gaping mouth, it won't suffice.
The pangs of hunger in her stomach's pit
No dishes can allay, nor cups her thirst.

[47] Nigellus here introduces a play on the Latin words *caput*, "head," and *capio*, "I take." Since the head, in this case, Rome, is so grasping, the author suggests that *caput* must be derived from *capio*.

Ah! nothing satisfies, nor all the world
  Can satisfy her needs increased by wealth.
Her greed condones each avaricious act,
  Each sin her love for money expiates.[48]
A bulging purse frees sinners, pardons sins,
  But empty has no hold on sinful man.
There's not a thing so hard, so grave, or bad,
  Which bulging money bags will not improve.
They legalize the things you want to do,
  And likewise make unlawful what's allowed.
He who alone his own wounds cannot see
  Corrects the faults of others, not his own.
It's thus a headache makes the body ache,
  And small amounts of poison taint large jars.
It's thus one spotted sheep defiles the flock,
  And massive things are spoiled by slight decay.
It's thus a stream, though far away, tastes like
  Its source, and from the fire smoke gets its scent.
From head disorders limbs have suffered pain,
  And children suffer from their father's sin.
Once righteousness had honor and esteem,
  But she who led the world now leads in vice.
Old Nestor's age would not suffice for me
  To tell each thing that ought to be recalled.
But to be brief and put in simple terms
  What people say (they often speak for God),
From crown of head clear down to sole of foot
  There's nothing left that's sane and sound, it seems.

[48] "Quicquid delirant caetera bursa luet." Cf. Horace *Epistulae* 1.2.14: "Quicquid delirant reges plectuntur Achivi."

He Speaks Also of Kings

SHOULD I unfold the life and ways of kings,
What might I find but cause of major grief?
Their life brings grief, their sway is harsh, their words
Are just like wind or fleeting breath of air.
Though man has been created to be like
The one who made from nothing all that is,
Yet kings have more concern for earthly things
And place a lesser value on mankind.
How often do they torture men and hang
Them on a cross for taking flesh of beasts?
What might Sicilian tyrants do more cruel
Than have men lose their lives for death to beast?
Who rule their kingdoms thus, who rule without
An oarsman, hardly bear the name of king.
Although their kingdoms are enslaved to them,
The people taxed, the country held in fear,
Yet these reign not, because to reign is not
To cause alarm, or rule with iron hand.
'Tis aptly said that such no rulers are,
But thieves who from their deeds derive their names.
They're unreliable, and he who trusts
In them is foolish and deceived, I think.
Their love is feigned, unfaithful is their way,
Their tongue is full of guile, bloodstained their hands.
They show respect for gifts, yet while they do,
They cast both law and justice far away.
For all the things they do or say or think,

All these, if you note well, are hints to give.
And if no gift attends or follows close,
As calf her dam, not far away but near,
You'll be no better off for title or
For habit, order, or for high repute.
Gifts frustrate kings, again by gifts they're saved.
Gifts grant us peace, gifts lead us on to war.
Gifts overthrow both bishops and the kings,
Gifts strengthen justice and destroy it too.
Gifts render eloquent the tongues of fools,
When gifts speak out all other things are still.
Gifts nullify the laws, revoke decrees,
They void our fathers' statutes by their power.
Gifts cover up the great men's faults, and hide
Their wicked deeds that they may prosper more.
Gifts—silence! lest perchance I should reveal
The secrets often found in holy hands.
Gifts show the wishes of the greedy mind,
Or what the sinful mind of man desires.
Gifts often show, if one may so believe,
What strength the sees without a bishop have.
Gifts prove a pleasing evil to the well,
A harmless virus; to the sick, death's mask.
Gifts always carry with them something else,
And citizens, though silent, make lament.
Gifts turn the scale of justice, and they shift
At once the balance toward the side they're on.
Gifts proffer cups with deadly poison filled,
They blind the eyes and cause one's steps to fail.
Love conquers all,[49] yet love by gifts is won,
And he who doubts this truth should weigh the facts.
Gifts stifle budding virtues, vex the saints,

[49] "Omnia vincit amor." Cf. Vergil *Eclogues* X. 69: "omnia vincit amor"; also *Ciris* 437: "omnia vicit amor."

They drag the wicked to the pit of hell.
Gifts spoil good character, and gifts received
For love lead to a loss of chastity.
Gifts elevate the bishops and the kings,
Gifts purchase crowns and cheer the hearts of dukes.
Gifts undermine strong towers and level hills,
They bring on death without the loss of blood.
When gifts precede, so many ills ensue;
When gifts arrive, a legal suit is near.
If presents cease, disputes and quarrels will cease,
And Mars will fall, to Venus be no friend.
If presents cease, the bloodless peaceful fall
Of Rome will come, impossible before.
If presents cease, the primate's robes will cost
Far less, and at a bargain will be bought.
If presents cease, the abbot's lengthy horns
Will be of lighter stuff and lower cost.
If presents cease, the monk recalled from court
Of king will tread the cloister's sacred walks.
If presents cease, the flock will rest beside
The shepherd; both will have a common mind.
If presents cease, perhaps into the cells
Of Cluny God will come again and stay.
If presents cease, the clergy will remove
A Judas and a Simon from their midst,
For often there are carried in their hands
Base things; with gifts their hands are often filled.
If presents cease, no longer will there be
For wretched mortals cries or grief or pain.
If presents cease, 'twill mean the end alike
Of Judas and of Simon, and their gold.
Gifts cause a king a tyrant to become,
And turn his hands against his fellow men.
The populace, unlettered, lawless, fierce

In manners, speeds to every act of sin.
The fierce, flesh-hungry, and bloodthirsty mob
Equips armed bands for every kind of crime.
That fear of God which once possessed both kings
And peoples is no more in name or deed.
What say you then? Since God is left behind,
No need for other god, but each his own.
The foolish people with their king have strayed;
The wether's feet one lot has made the same.
Behold such vast destruction plagues the world,
The days that lie ahead are grimmer still.
Nor have we one to save or ransom us,
But hope of our salvation is all gone.

Concerning Evil Bishops and Pastors

IT'S MY belief no others in these times
Are more at fault than holy bishops are.
Though shepherds are obliged to keep the law,
Yet they neglect to do the things they teach.
They should be called, instead of shepherds, thieves,
With God my witness, who thus speaks of them:
Great numbers of false prophets shall arise
And practice on the many their deceits.
Though in sheep's clothing they may come, yet know
That they are hungry wolves athirst for blood.
They get their name of shepherds of the flock
By feeding on the sheep, not feeding them.
The name officially conferred they've changed,
So they're not shepherds of the flocks, but wolves.
The three main creatures which stay near the fold

Are these (and each desires the nearest place):
The shepherd, who is first, the hireling next,
And then the lurking wolf, which makes the third.
The first one freely feeds his sheep, the next
For pay, the third to ravish and destroy.
The shepherds then, in doing nothing free,
Bear no resemblance to the first at all.
And when they feed them through their love for gain,
They're like the second and they share his lot.
They have no things in common with the first,
But with the second much, there is no doubt.
Then like the wolves, they eat their flesh and blood,
And tear the young from out their mother's womb.
Like wolves the shepherds of today come forth
Each one to kill the choicest of their flock.
And what they see to break they break, what's weak
They spurn; the fat they eat, the lean they scorn.
Sheep's wool and milk they seek, and they disperse,
Subdue, and wretchedly destroy the flock.
Nor does it sate them just to loot the fold,
To rout the flocks and ravage for themselves,
Unless 'tis theirs to drink the tepid blood,
And theirs to thrust a hand against their throats.
How in these times may shepherds come into
The folds, and who shall lead, and by what way?
And, having come, what may they do, how serve
With zeal the flock and not the market place?
If we examine closely shepherds' lives,
We'll find they do a host of shameful things.
When placed before a glass where they can see
Themselves on every side, they close their eyes.
They do not scorn concern for flesh and blood;
Their chief concern is for themselves and theirs.
Whom they should most of all give preference to,

He scarce can hide himself in distant lands.
He's hot, cold, needy, thirsty, hungry, bruised,
Who gave whence each his hunger might relieve.
So many dishes grace the bishop's board,
So many liveried servants, countless cups,
So many lords and servants everywhere,
So many boys, both standing and astir;
So many youths adorned and wreathed with crowns,
So many elders versed in deeds of old.
They run about like soldiers in the camp
Of kings, at ease in mind and body free,
And while the master drinks, one lifts his arms
And falling on his knees pays his respects.
The house is bright, the gold and silver shine
On which is served to God the sacred host.
The things provided for the bishop's use
Are not composed of cheap or shabby wares.
They know not what is proper who present
The bishop cups of gold, but tin to God.
The bishop lies on gold and silver beds,
The house of God scarce has the use of clay.
Dilapidated churches you can see,
Their altars bare, their crosses all despoiled.
In splendor shines the bishop's columned hall,
Of marble columns both inside and out.
So many are the bishop's tunics, gowns,
And hoods, that he can't keep account of them.
The bishop's fingers hold sufficient gold
And gems for seven men of wealth to own.
The bishop likes to study marks, not Mark,
And is more fond of lucre than of Luke.
Day after day, since God long pays no heed,
They thus hold things in trust, and not as gifts.

To whom and how they give the keeping of
 Their souls, the most important thing, is clear.
Before a child knows how to speak the names
 Of parents, or can stand or walk alone,
He takes the church's keys and guides our souls,
 While still a baby crying in his crib.[50]
Now what will Peter say when little Bob
 Or Willie guides and leads me to the stars?
To Peter, not a child, God gave the keys
 And bade him guide the church's destiny.
O Lord, do not entrust me to the care
 Of Bob or Willie; have concern for me!
Nor shall I yield, although perhaps, may heaven
 Forbid, the bishop force and worry me.
Disturbed by this I'll first appeal; I'll go
 Alone on foot to Peter and his heir.
With nurse will come to Rome our little Bob;
 He'll come in new or covered bassinet.
And Willie wrapped in bag will also come
 To Rome, and he'll be seen by all the court.
And they'll not go with empty hands or waist,
 But like a woman soon to bear a child.
More closely will they draw and come to see
 The tomb which ever wide and empty stands.
Whate'er is needed by this tender age
 Is furnished, thanks to those who share the way.
Their father and their guide will give to them
 Such years as they will need, if need there be.
Did he who gave us churches not confer
 What years and manners such as these should have?
Who gave one thing could likewise grant the rest,

[50] Nigellus again criticizes the practice of conferring the office of bishop on small children in his *Contra curiales*, p. 163.

By equal law each thing is granted him.
Young beardless boys made pastors, we have seen,
Of churches, and made holy bishops too.
Thus spoke a man of one about to be
A bishop, when the king incited him:
'A child this is; not yet can we discern
The sex, and soon a bishop he will be!'
To such belongs control of church and men,
And these are keepers of the Master's house.
The church of Christ on pillars such as these
Will tumble down, destroyed before its time.
If you ask why the bishop runs through town
With dogs, he plans to go into the woods,
That with his falcons he may fowl, or cast
His hook to fish and catch a pike or gar.
He spies a heron on the river bank,
The bishop runs from town to throw his hawk.
He spends more time in woods than sacred place,
And values dogma less than sound of dogs.
He's troubled more when dogs are lost or when
A bird is hurt than when a cleric dies.
He often spends all day on idle things;
An hour suffices for his holy tasks.
A long-drawn-out dispute seems short to him,
Yet what concerns the Lord much boredom brings.
A dog more closely crowds upon the hare,
As quietly he consummates his task,
Than bishops, when the time and hour require,
Perform for God the service that is due.
If any suffer for the Lord, and that
Because of paltry things and brief delights,
Let no one doubt that they, while in the flesh,
Are living saints of God and martyr-like.

Yet these are not ashamed to suffer thus
And don't regret the yoke, though often hard,
Since God for whom they suffer for an hour
These ills will give to them their due rewards.
I wonder more why abbots, priors, and
The sacred convents whom their rules restrict,
Should disregard the rules they pledged to keep
And which the holy fathers set for them,
And backward turn and eat what they spewed out,
As dogs return to vomit, swine to mud.
Externally they bear a righteous front;
Inside they have a heart that's filled with guile.
Who follow in the steps of Benedict,
Bernard, or Augustine, with lighter yoke,
They all are thieves in spite of sacred stamp
By which they come and magnify the Lord.
Trust not their words, trust not their spotless robes,
For scarcely may you trust the things they've done.
They have the gentle voice of Jacob, but
The rest is Esau's—arms, the neck, and hands.
They've gone again to Egypt which they left,
They think it sweet to be in Pharoah's hands.
They rush headlong to pleasures of the flesh;
They're drawn by flesh into the pit of death.
They're goaded on by envy and distraught
By surges of ambition, chiefly three.
They first aspire to rise, and next there comes
Unbounded love and grave concern for kin.
These fire their hearts, for these they strive,
These hold command, with God thrust far behind.
And next, that they have wealth, they labor hard
That they hold briefly what they've long acquired.

Concerning the Religious in General

ONCE poverty and lack of transient things
Gave crowning glory to religious men.
But now if they have not what damns their souls,
Great riches, pasturelands, and fields and flocks,
They think they're wretched, for they now believe
It's bad to have the world regard them poor.
Yet Christ was poor, whom they so imitate
That they thus follow him in name not deed.
They wish to seem what they don't wish to be,
As if to hide from God who sees all things.
When they had put all things aside, they aimed
To walk the paths of Christ in word not deed.
They spurned the world, but on such terms
That once rejected it be ever near.
That they not lack they spurned all things, for one
Who lightly sows to harvest much succeeds.
A wolf dressed like a sheep is not suspect,
Yet must be feared by reason of his teeth.
The fox walks straight along with gentle mien,
There is upon her face no hint of guile,
But by her craft she conquers many beasts,
And always slyly waits within her lair.
The higher up the hawk ascends the sky,
The better does he catch from there his prey.
The ram retreats to strike with greater force,
With lowered head receives and deals the blows.
Thus are those acting whom the world today
Regards and calls devout and holy men.

Concerning these, though much is left to say,
I say no more for fear I be rebuked.

Brunellus on the Laity, as Taught on a Certain Occasion beneath an Oak by the Chattering Birds

BUT WHAT about the laity? If I
Make public what I've learned, I'll be disliked.
Not long ago while fleeing out of town,
Escaping from Bernardus, fearing blows,
I searched for hiding places which that lord
Of mine, that wretched scoundrel, could not reach.
But he kept searching every hidden place
With pack of hounds and fiercely threatened me.
He swore an oath that if I should be found
I'd suffer for that journey for all time,
And he would write 'deserter' on my back,
So that I'd never fail to recollect.
At times while I was snoring he was pleased
To think he'd catch me and would come real close,
But I'd be wide awake and jump right up,
And laugh at Sir Bernardus chasing me.
I'd often kick him with my foot and throw
Him on his rear, and make him snore like me.
'Twas summer, I was tired, I chose to stretch
On pleasant turf beneath a shady oak.
A poem which I had just composed I tried
To put in writing; safe was I from foe.
Both time and place for study were just right,
And since I wished to write, they urged me on.

My stomach was well filled, my feet were tired,
The page was ready to record my song.
And as I took my slender iron-tipped pen,
And on the parchment was about to write,
I heard nearby some kind of chattering,
I heard the sound beneath the lofty oak.
So many soaring birds filled all the place
That though the woods was large, it soon was thronged.
The forest thundered, and the sound of birds
Spread through the air toward heaven, shaking earth.[51]
As for myself, I feared and in dismay
Began to cry: 'What means this strange event?'
The sight with terror filled me, nor 'twas strange,
Because I did not know the place and was alone.
I listened closely and remained stock-still,
Prostrate I lay, and made no sound at all.
I softly smote my breast and pardon sought,
In thought not word, and with a humble heart.
It almost happened through excessive fear
My stomach was upset, and I was shocked.
'Tis true that heavy winds and fickle storms,
When rain is light, will all at once grow calm.
The noise abated, silence reigned at last,
And all of them became at once quite still.
From out the company, when all was calm,
A single raven rose and thus began:

The Raven's Opinion

" 'BELOVED BRETHREN and distinguished sirs,

[51] R. R. Raymo, "*Parlement of Foules* 309–15," *Modern Language Notes,* LXXI (March, 1956), 159, shows that this passage provided Chaucer with material for his *Parlement of Foules.*

Please listen, for I have some things to say.
It's not that I'm so good or holier
Than thou, for I have come from lowly stock,
Or that I'm wiser, since, as David says,
The source of wisdom is the fear of God,
But since I'm more advanced in years than all
Of you, as length of years will bear me out.
Wise Noah brought me to the ark when seven,
While rolling waves concealed the mountain tops.
Since then how many years have flown along,
Unknown to most of you, but very good!
As I reflect on them and calmly muse,
How good they then did seem, how filled with good!
Respecting present things I sadly fear
The day of doom is drawing swiftly near.
Indeed all glory of the aging world
Is gone, all things have lost their former state.
The stars, the earth, the sea, are different;
They've ceased to be controlled by ancient laws.
And if we know the present by the past,
No members to their bodies will hold fast.
The order of the ages has been changed,[52]
The world is not the same as yesterday.
The universe to nature is opposed,
Both day and night will fail to keep their course.
Hence sickness slays the people, pestilence
The beasts, and birds succumb before their time.
The case is clear, for all of us have sinned.
Yet none is left who thinks that he has erred.
They all are righteous, none has sinned or erred;
No sin did I commit, nor you nor he.

[52] "Ordine transposito variantur saecula quaeque / Alter et est hodie quam fuit orbis heri." There is some suggestion here of Vergil's famous line, "magnus ab integro saeclorum nascitur ordo," *Eclogues* 4.5.

I've tasted nothing, all the servants said,
And yet it's clear the wine has been consumed.[53]
Suppose one thought to grieve about his sins
And wished to offer penance for misdeeds,
Suppose he wanted to confess his sins
With contrite heart, in tears and deep remorse,
To whom of us he safely could confess
His secret sins, I surely do not know.
One thousand each of brothers, sisters, and
Of fathers too, and mothers let me have,
Of these there'll be one conscious of my sin,
And he alone will be my counselor.
If raven to a raven I confess,
He'll tell the crow, who'll caw through all the town.
If I confess to cock, the hen (O Lord
Forbid!) will to her young ones point me out.
To speak or not to speak means death to me;
Two things sore vex me, fear and sense of shame.
The tongues of priests, since they tell secret things,
Compel us to confess unwillingly.
There is no greater evil than to have
The very one who knows reveal your sins.
Who falls upon his sword must once be mourned,
But he forever whom his own tongue slays.
You then who either do or notice done
The countless things which you should disregard,
With me your model learn, I pray, what you
Should say to whom, lest later you repent.
Learn who seeks out and often wants to help
The fallen and to cheer the comfortless,
Who crushes not the freshly broken reed,
But bandages, restores, and long sustains;

[53] I am following Sedgwick's suggestion (*Speculum*, V, 293) that Wright's reading *canem* ("dog") be emended to *cadum* ("wine-jar" or "wine").

Who pardons those who've erred, and pities them,
And fears that he could suffer such a fate;
Who always thinks how weak is flesh, and how
Disposed it is to fail when life is hard;
Who, deeply wounded, flares not up in wrath
Nor seeks quick vengeance on his enemy,
But good and patient he himself forgives,
Unasked, another's every sinful act;
Who, self-sufficient and despising wealth
And glory, takes no gifts, though very poor,
Whom neither love of world nor parents' cause
Disturbs, nor gifts (sweet evil) overthrow;
Who neither loves nor does what should be blamed,
With whom to live or not is all the same,
If such confessor on the earth be found,
I'll give him full confession, I confess.
I give you warning, since my babbling tongue
Once caused me mischief, as my color shows.
While silent I was loved, but my loose tongue
Made me despised. Alas! O busy tongue.
How rich was I and blessed with many friends
Until my tongue deprived me of myself.
Had I been still as bust of Mars,[54] I'd be
Well off and keep the color which I had.
What once I lost is gone beyond reprieve;
I fret and pine in everlasting grief.
Fame, glory, honor, have been lost through talk;
My tongue undisciplined took all away.
Loose tongue, I grant, did harm, for had I held
My peace, my native gifts would all be left.
Of all the birds I was the prettiest,
And was alone most pleasing of them all.

[54] "Felix si statua Martis taciturnior essem." Cf. Horace *Epistulae* 2.2.83: "statua taciturnius essem."

My voice, once clear, is coarse and almost gone,
   My wretched self is witness to my sin.
My fame would still be left, had I been still,
   But since I spoke, it suddenly was ruined.
With thought of me then hold your babbling tongue,
   Lest, getting careless, it may ruin you.
Few things I'll say, in deference to my lords,
   But true, if there be any place for truth.
We must not wonder, though adversities
   May often come and many ills befall.
We are ourselves the cause, if cause we seek,
   For falling thus among these many ills.
We ravage crops, we steal and take away
   The seeds, and devastate the copious fields.
That they not grow we rend with claws and beaks
   Those freshly bursting blossoms on the trees.
We thrive on constant slaying and rapine,
   A wanton army, indisposed to good.
The peasant thinks he safely stored the grain
   Which he has threshed behind a bolted door,
But there arrives the cock with all his flock,
   And takes as much as he desires, and flees.
To steal is not enough, unless what's left
   He scatter with his feet about the house.
The falcon and the hawk, when they are fed,
   Fly off and will not to their master come.
They disregard and shun the youngster's tears
   Who calls and stretches out his arms to them.
They spurn that friendly hand which often fed
   And carried them, as if a stranger's hand.
Though certain risks of death assail the boys,
   No pity do they have for them at all.
And while they chase them through deserted tracts,
   They often lose their lives in woods or streams.

'Tis thus we bless our friends for kindly acts,
And for a good turn often do a wrong.
The parrot, when the mistress reappears,
Betrays the girls by telling what they said.
What they in private and in fear have said
To cautious ears, and with no witness near,
The parrot tells, he who can't talk, nor yet
Keep still, because he has too large a mouth.
The girls then often feed birds aconite
That they may die before they learn to speak.
And there are other birds which all their lives
Inhabit sacred cloisters of the monks.
How many dangers through the year these bring
To them, I cannot now enumerate.
Hence people hate us with a deadly hate,
And often pay us back with like return.
Hence every day they lie in wait for us,
Their hostile hands remove us from their midst.
Hence they set snares for us and stretch out nets,
Hence stone or winged arrow brings us death.
Our feet and wings are held by sticky lime,
And we're deceived sometimes by pretty sounds.
Not undeservedly we bear these ills,
And pay perforce great penalties we've earned.'

The Cock's Reply

"THE COCK then answered him and said, 'My friend,
Refrain from speaking, that's enough, be still.
Your words lack understanding and good sense,
Nor does your speech take hold upon my heart.
In your discourse you're too diffuse, and like
Old men you're full of words and lack good sense.
'Tis said your tongue once made you suffer harm,

Nor has it yet brought any help to you.
Much you have seen as one who's lived for years
And backward looked upon the ages past.
It's not surprising then if senile now
You ramble, and your mind is void of wit.
As for the old man, as his years increase
His powers of mind, and body too, decrease.
The senses in the dotard are eclipsed,
For not by reason is he ruled, but chance.
A brain disordered weakens all the parts;
The eyes, the tongue, and hands are made to fail.
You then rebuke and censure us as wrong,
As though yourself the just and blameless one.
You're just and surely known through all the world,
As Noah's ark, not I, so well proclaims.
Did Noah not expect you to return?
A carcass was the thing that held you back.
Your gluttony and vain desires prevailed
On you to be a traitor to your lord.
If you will weigh your life and ways and deeds,
You'll find none like you in the whole wide world.
From tender years you've learned to be a fraud,
You can't now change your ways when you are old.
Through gluttony and taste for carrion,
'Tis known that you from youth have broken faith.
For kindness it was wrong to leave your lord,
'Twas evil to desert at such a time.
That all things else not be disgraced through you,
You left, a traitor, never to return.
Oh how unlike am I to you and how
Opposed my ways and actions are to yours!
For whatsoever master I have found,
I've always been his true and loyal slave.
And from me too have come throughout the year

Advantages for poor and rich alike.
From me are given feathers which provide
Soft beds to rest the weary bones each day.
From me the feeble and the hopeless get
For wholesome food the pullet's tender flesh.
The fattened capon also from our breed
Provides a richer food for healthy folk.
The source from which fat chicks and eggs derive,
Originating first from out our loins.[55]
More wide awake at night than are the guards
I signify by sound of voice the hours.
I waken lazy boys and those asleep,
Nor do I let them lie in cozy beds.
And many more advantages I give
Which I need not enumerate right now.
Yet not ungrateful nor remiss to those
Whom fate has made my masters would I be.
For should I wish to show ill will and not
(O heaven, forbid!) put curbs upon my lips,
I'd see that many were removed and got
Their due deserts in jail and on the cross.
For many wrongs in dark of night are done,
And even darker than dark night itself.
If what the master does and servant says
The market-places knew as well as I,
On crosses you would often see at dawn
Suspended many men and women too.
What country bumpkin whispers to his wife
He cannot, though he wish, conceal from me.
The gods wish not that I should harm a soul
Or by my voice betray another's crime.
Forbid that I be like the raven who

[55] There seems to be some difficulty in the text here, and I have therefore been obliged to make some slight readjustment in the lines.

Betrayed the one who was his chief support!
I pray that Jove may hurl his bolts on me
And that the earth devour my tribe and me,
May ocean waves engulf me and I fall
Headlong into the depths of hell below,
Ere I shall any secret trusts betray,
Or evil word find exit from my mouth.
Unless we praise, commend, respect, and love
The men with whom we must associate,
We are but fools and like a statue have
No human sense, but sentient bodies have.
My thought is this, my true opinion this,
I say these things on my account alone.
The falcon, hawk, and others may speak out,
They're old enough and have a mouth and voice.
To farmers I've been forced to be attached
By lot and live far off from palaces.
Unless therefore by force I'm brought to court,
A wooden tail will then appear on me.
Because the hawk and falcon are at court,
They share the councils both of kings and dukes.
In private cells apart from noisy throngs
And far from crowds it's granted them to live.
High in the prince's chamber they are placed
That they may notice many things from there.
The farthest corner of the room, a place
For doing secret things, is given them.
There evil plans are frequently devised,
There wicked deeds are very often tried.
It's very pleasing both to boys and girls
To have a daytime corner dark as night.
Here stealth is often hidden which is forced
To come into the open all too soon.
Here too confession oft is made, but not,

I think, to any priest nor to be made.
Here poisons lie concealed, acquired by toil,
And here the dreadful spells of step-dames thrive.
This place requires a boy to come in dead
Of night to see the hawk when he is ill.
This place cures those with asthma and it helps
The epileptics; day is turned to night.
These places then are such as hate the lamps,
And thanks to vice obtain rewards of praise.
The falcon and the hawk know well the spots
Of evil deeds, and know whereof I speak.
These two indeed, should they desire to tell
The truth, strange things could surely say, but true.'

The Reply of the Falcon

"THE FALCON, having heard these words, replied:
' 'Tis wrong, I think, one's duty to neglect.
Of noble birth are we, of high descent,
And it's improper that we speak base words.
No matter what the course the boys pursue,
We should, I think, be ever tolerant.
If boys amuse themselves, yet harm us not,
Why bother them? Away, such foolishness!
Let's not betray those who've been free to let
Us know their secrets, and to whom we're bound.
The boys feed us, they carry us around,
They also take the greatest care of us.
Informers are deservedly despised;
And to accuse a friend of wrong is wrong.
When eyes that see unusual sights don't speak,
Why speaks the prattling tongue when it sees not?
Forbid it that the mouth thus serve the eyes,
That it should be the cause of sin for both.

If you indulge an evil mouth, though one,
What need for you to keep both eyes closed tight?
Let then the evil mouth be closed until
The ears, eyes, feet, or hands express themselves.
The tongue can then its utter silence break
When it has heard the other members speak.
Free speech will then be rendered unto all
Of whom at present question has been raised,
But with this added that if they can't be
On guard, it hold its peace, if possible.
For of the many heinous sins on earth
There's none of them that can be worse than this.'
While thus the falcon spoke, sleep closed my eyes,
And I began to snore, as is my wont.
The sound disturbed the birds, and swifter than
A springtime breeze they quickly flew away.

Brunellus Also Discourses on the Inconstancy of the World

WHILE often thus, my teacher, I reflect,
I'm very sorely troubled and perplexed.
And since there's nothing constant in the world,
Who'd claim the constancy of anything?
Behold! I once was young and labored hard,
Was strong, but now I'm old and broken down.
And though I know I've lived one hundred years,
It seems to me it's hardly been three days.
All things attest the tomb alone remains,
And if a day is left, it's short, I think.
To want long life is but a wish—forbid!

To live in sin—let life therefore be brief.
Since from his mother's womb the child has not
One sinless day, why think that I shall have,
Who always sin and am a wretched fool,
Not fearing man and disregarding God.
Not blindly do I sin, but knowingly
And shrewdly I a churchman do bad deeds.
And while I think and say and do bad things,
Next day my wretched self does even worse.
I'm worse progressively from day to day,
For I go on, inclined to every sin.

Here Again Brunellus Proposes To Devote Himself to His Religion

"HENCE I INTEND to enter holy life,
In which I'll be head teacher and in charge.
This with God's help I'll enter soon, because
I'm troubled by the shortness of my life.
I'll therefore piously proceed to Rome
That this new order may be formed by me.
Not stern in this, I think, will be the Pope,
The court will gladly give its full assent.

Brunellus Advises Galienus To Enter the Same Order with Him

"AND IF, SIR, you should wish to change your life,
I warn you not to plan to go too far,
Nor for your soul's salvation be ashamed

A teacher to your pupil to submit.
It often happens in religious life
That he who's greater serves, the weaker rules.
Good things live under bad—the rose, wild nard,
And lilies, thistles—be not moved by this.
Slave overwhelms the free, the fool the wise,
Unjust the just, and darksome night the sky.
Men often magnify with greater zeal
What they find more abased or of less worth.
They illustrate by deed a story which
My mother often used to tell to me.

He Introduces Here the Story of the Three Goddesses Who Are Called the Parcae and Who Are Believed To Spin the Threads of Destiny

THERE went three sisters to relieve men's cares,
Whom we declare the goddesses of fate.
These three had one concern and one desire,
To bring to nature's flaws good health and strength.
Where nature grudged or lavishly bestowed
They very much desired to make amends.
As they once walked along they found a girl,
Fit maid for Jove, at foot of shaded hill,
A pretty miss indeed, of noble birth,
Whose lovely face revealed nobility.
For her the gods, had she been known to them,
Great wars against almighty Jove had waged.
And Jove himself, to gain the maid, would choose
To be exiled from heaven seven years.

No more could nature have increased her charms
 If she had sworn by Styx or were Jove's child.
Yet she complained, and shed so many tears
 That often she made wet the earth around.
She slashed her face and then she beat her breast,
 She scratched her cheeks, and gave herself no rest.
Two sisters then on seeing her desired
 To offer help, if it were granted them.
The twain besought and begged the one in charge
 At least to make the evil less severe.
But she, contrariwise, by look and words
 Refused, and to their pleas a deaf ear turned.
But not to frighten or to worry them
 Or seem to hold them in contempt, she said:
'We've come, you know, we three to see the world,
 To be of help, but only where there's need.
No need has she, for nature's blessed the girl
 As much as it could do or dared to do.
Since she has noble birth and special charms,
 That ought to satisfy both her and us.
Perhaps if we should offer her our help,
 She soon would suffer evils worse than these.
Abundance has harmed many; better to
 Have lived contented with a little than
To seek beyond their powers another's lot,
 And to enjoy what's neither right nor just.
Things unattended by the needs and aims
 Of nature soon succumb and quickly fail.
Not long does nature let strange things be joined,
 Nor is she forced; by reason she is ruled.'
She spoke; they left the sad and weeping girl,
 Her state unchanged from what it was before.
The blazing sun beat on them as they walked,
 And heat and toil both greatly slowed their steps.

And so they turned into the wood nearby
To quench their thirst with water from a spring.
No sooner had they reached the wood, when lo!
There lay a lovely girl upon a couch.
When to the goddesses she fain would rise,
She could not for her massive feet forbade.
What she could do she did—stretched forth her hands,
And in a gracious manner greeted them.
She told them of a forest path which led
To where there flowed a fountain clear and fresh.
She also added, 'I would go with you
So gladly, if my fates would but allow,
But due to weight and pain in feet and hips,
I'm held forever fixed upon my couch.
I came here for the shade supported by
Another, not myself, and I'm alone.'
When they had heard these words two sisters wept,
And weeping sought to melt their mistress' heart.
And though they cried and begged unceasingly
That she restore to her the power to walk,
To stand on her own feet and to return
In health, no longer to remain in bed,
She was unmoved, and for her sisters felt
No sympathy, but leaving said to them:
'You're moved, I see, by pity, as is right,
And now I know full well your heart's desires.
They're tender but improper ones, for those
Not based on reason should be cast aside.
Behold the circumstances of that girl
For whom you've earnestly besought my help.
She's kept from walking by her heavy feet;
This constant weight confines her to her bed.
Her other members have their special gifts,
If you consider well, in ample store.

Both keen of mind and strong in voice is she,
  A life of study occupies her time.
Such grace of hands on her has been bestowed
  That hardly anyone can rival her.
She has, this fair-cheeked girl, three winning charms;
  Her breast and voice and hands suffice for her.
Alone, with hands alone, she could be match
  For ten girls—why is she alone prostrate?
No need is there for her to have our gifts
  Since she by mighty nature is so blessed.
If she is weak from loss of feet alone,
  She's strong in three or even four respects.'
On hearing this the sisters dried their tears
  And started on the journey which remained.
Approach of evening bade the goddesses
  Seek rest and comfort from the toils of day.
And when they now were settled for the night
  Inside the entrance to a town nearby,
There came into the road a peasant girl
  To defecate; no fear or shame had she.
She raised her dress behind and drew it back,
  She flexed her knees and squatted on the ground.
One hand held grass, the other piece of bread,
  Both hands performed due service for the wench.
But she cared not for people as they passed,
  Nor for the holy deities nearby.
Two blushed to see the sight, and hid their face,
  And having veiled themselves they took to flight.
The third stood still and calling back the two,
  She said, 'What means your flight? Remain, I pray.
What troubles you? No idle dream you see;
  You've seen just now the other side of life.
Nought better had she than what she revealed
  To us and showed in her simplicity.

Had nature given the wretch a better lot,
  Not thus would the new moon have shown its horns.
There's need, there's need to pity, not to grudge,
  To offer her our help right speedily.
There's need, there's need that we be generous,
  And give abundant gifts with lavish hand.
To her, rich nature has left nought at all;
  She's poor; there's need that we bring help to her.
She'll never by her strength alone arise,
  She needs an act of mercy to be saved.
If once she fall, there's none to lift her up,
  Nor has she any strength to lift herself.
Poor helpless creature nature thus has made,
  Because instilled in her there is no grace.
That man by nature not by fortune blessed,
  'Tis he who by his talents has true wealth.
'Gainst nature not one thing can we implant,
  Nor wish we to, except by reason's laws.
Inside and out our nature stays the same
  As was before, but circumstances change.
We wish not to debase what nature does;
  It can't be done, and we should do no wrong.
Material objects can accumulate,
  But nature keeps her own identity.
Let us therefore to whom this power is given
  Present huge gifts with an unstinting hand,
Wealth, riches, money, offices, and farms,
  Estates in mountains, pasture lands, and herds,
And let us make her mistress of this town,
  That nothing fail her which is ours to give.'

Narration of Brunellus

MY MOTHER often told me things like this,
To mention which I surely have no shame.
So many things like this which make no sense
Inside religious orders oft transpire.
This situation, sir, so frequently
Exists, and you must often bear it too.
It's common, many specialize in this,
The bishop often has this evil trait.
It's what produces lack of harmony,
And separates the pastor from his flock.
Those things which bring destruction to the church
Are slander, schisms, losses, scorn, and guile.
Though virtue can submit to most of these,
She cannot close her eyes to private scorn.
There's nothing wounds or burns the mind as much
As when the gifts of virtue are despised.
Though other wrongs are oft forgot in time,
This hardly ever will desert the mind.
This kind of wound, though treated, gets much worse,
And causes greater pain as time goes on.
It causes venom to infect the brain,
For which no antidote will bring relief.
He who can bear with patience private scorn
Can easily sustain all other things.
Though patience wins in nearly every case,
Beneath this heavy load it meets defeat."

Brunellus Fears Coming Evil from a Sign Which Appears to Him

FROM his left nostril as he thus did speak
Cold blood burst quickly forth, but not for long.
On seeing this at once Brunellus said:
"This sign portends some evil thing for me.
Some time ago, the night before the dogs
Tore off my tail, this very thing occurred.
Lord, grant me now good luck and rescue me
From evil! Let this blood bring good, I pray!
Each morn let running hare or limping goat
Present themselves to those who wish me harm!
I pray the owl may be the bird which first
Comes forth at dawn to bring my foes bad luck!
Let barefoot woman spinning black sheep's wool
On distaff made of yew encounter them!
Let bald old man at morning meet my foe,
And stooped old hag with rumpled hair at night!
Let monk of both the colors meet them too,
His head bowed down, and saying not a word!
Let priest in anger greet them as he leaves
A pauper's or a widow's last sad rites!
Let toad athwart their pathway on the left
Be first to meet them and to hop along!
And let the rustic who is often called,
And is, the deadliest pest assail my foes!"

How Bernardus, Brunellus' Master, Overtakes Him

E'D SCARCELY spoken thus when lo! there came
A cruel rogue who closed the door and said:
"O you, Brunellus, sought for many years,
By merest chance, I've found you now at last!
Behold your lord, Brunellus, yes indeed,
Whom you'll repay for all your past misdeeds!
Come here to me, for back to town you'll go
Which you by night deserted, working guile.
'Twas New Year's and the holy Sabbath day,
When you dared perpetrate this sinful act.
And if I quite remember, since you left
Some twenty years or twenty-five have passed.
Now broken by old age, with strength and health
Of body gone, you claim that you are old.
Forbid that you do anything again
By force or willingly that's free from toil.
You'll only have to serve my daily needs,
Do nothing save to bring a little wood,
Two copper baskets, and two sacks of meal,
And me besides—I'm not a heavy load."
He spoke and having haltered him, he led
Him forth as usual, prodding with a club.
And lest again he think to run away,
And rob his master of his services,
Bernardus cut clear off his ears, that he
Become more cautious by this cautery.

Observation of Brunellus

TO HIM THEN SAID BRUNELLUS: "Now I know
As true, because my head is like my tail,
The words the true and holy prophet said
To me at Paris once, while I was there.
He said the future's often like the past,
And much foretold I shall not mention now.
Indeed what things he said would come to pass
Have turned out just as he had said they would.
He also often spoke to me of you,
And many things foretold which I recall.
For nothing that he said has turned out false,
Nor were his words without significance.
With foresight he explained to me my fate,
And told me what he saw would come to me.
He prophesied I'd have five strokes of luck,
When five adversities had been endured.
Of those five blows of fortune which that seer
Of mine predicted four have now been met.
The fifth which I must suffer, and the last,
Remains for me, and it I now endure.
This last will be the last of all my woes;
The end and limit of those toils is near.
The last will end my troubles and will bring
Relief, and I shall have my prayers fulfilled.
For having once survived the five bad fates,
There soon will come to me as many good.
This wicked rustic, as he was before
My source of grief, will also be its end.
The source of all my troubles from a child
Has stemmed from this contrary quarrelsome knave.
Through him the status of my life was changed,

For all the time before I had been free.
He slyly forced me first to be his slave,
And made me bear a heavy load of work.
But my discomfiture will have its end
In converse order whence it first began.
Not I, believe me, but that spiteful knave
It was who set the pattern for these ills.
To render him his just deserts will be
Not wrong, but certainly a righteous act.
An ancient law demands a limb for limb,
And tooth for tooth, and even foot for foot.
Thus shall I treat Bernardus, if the fates
(Alas!) chance not to thwart what I desire.
If only he could see the things to come,
How I Brunellus shall receive great fame,
Not thus would he be plying me with whips,
Nor strike me with a club nor spur my side.
But rather he would fall upon the ground,
And lying low adore my tracks, and pray."
Thus musing to himself Brunellus came
Into Cremona, subject to his lord.

On the Experience of Bernardus and Dryanus[56]

MEANWHILE Bernardus saw a thing take place
Worth telling and in all Cremona known.
The oft-repeated tale does not let things
Grow old, although the years may come and go.

[56] J. H. Mozley (*Speculum* V, 260 f.) connects the Dryanus story with the time of Richard I, who reigned from 1189 to 1199, and to whom this was a favorite tale. "Originally Sanskrit, it seems to have been localized in Italy," he says, "before 1195: under that year it is told by Matthew Paris as an apologue often repeated by Richard I."

The oft-repeated tale restores to life
　The face of things, and renders old things new.
The oft-repeated tale of former deeds
　Brings back time's losses and the dying age.
Three boys had poor Bernardus to support,
　A fourth care was his wife, a fifth his ass.
The ass, though worth a paltry sum indeed,
　Was all the substance of that house and lord.
His job of hauling loads of wood to town
　Remained the sole support of all that house.
Now once Bernardus chanced to go from town
　Into the woods accompanied by his ass,
And there he heard, it seemed, inside a cave
　The sound of human voice and calls for help.
Bernardus stood and was at first dismayed,
　And then across his brow he crossed himself.
And noting well the nature of the cry, he found
　It had the pitch and sound of human voice.
Perceiving where the voice first reached the air
　And sounded clearer, there he made his way,
And so the voice, in speech articulate,
　At last gave meaning to the words expressed.
A great and noble man he chanced to be,
　Dryanus of Cremona, wealthy too.
He often with his hounds pursued wild game,
　But now he chanced to fall into a pit.
The hole on top was small, but underneath,
　The pit was wide and very deep and dark.
Into it also fell a lion, ape,
　And snake, with which Dryanus made a fourth.
When these had slipped into the pit, he too
　Slipped in, became their guest and doubtful friend.
Each one was frightened and through fear was still,
　Save him whose voice because of fear grew loud.

Four days had passed since he with mournful cry
  Had begged without success that he be saved.
Aroused by this Bernardus came and asked
  That he tell who he was and whence he came.
" 'Tis poor Dryanus that you hear," said he,
  "Who until now was mayor of my town.
Whoe'er you be, come near and kindly help
  This wretched man, and win a great reward.
All that I have of worldly goods I'll share
  With you—I'll give you half as your reward.
And that you may be reassured, I swear
  To you by all that's holy to the gods."
On hearing this the rustic longed for gain
  And hurried to the cave to rescue him.
An osier rope he lowered, but the ape
  Caught hold of it, climbed out, and sped away.
Bernardus thought the devil had deceived
  Him thus, and said, "Oh, what a fool am I!
Look what came out! Dryanus should have come,
  But through the devil's wiles a mighty ape.
The place is haunted; 'twas the devil's voice
  And evil powers by which I've been deceived.
I'll then return, for I have crossed myself
  That nothing evil may do harm to me."
As he was speaking thus, Dryanus cried
  Aloud, adjured him, and renewed his pleas.
And lest he think himself deceived by ghosts,
  He named the saints and uttered holy words.
Dryanus begged that he again let down
  The rope, lest those rewards be lost by him.
A love for gain won out; again the rope
  He lowered, but in vain—his hopes were crushed.
Upon that rope there issued forth the snake,
  A fearsome sight, and quickly disappeared.

On seeing this Bernardus said: "Your gifts,
Dryanus, I refuse, keep all, away!
This place is filled with devils, ghosts spring forth
From earth, and from her bosom monsters rise."
He would have fled, no thought of coming back,
Had he not been drawn back by hope of gain.
His heart was ruled by hope and lust for wealth,
As with rich promises Dryanus prayed.
Just as before, though poor Dryanus stayed,
The lion climbed out upon the lowered ropes,
And this before Dryanus could observe
Or grasp with ready hand the dangling cords.
Bernardus, when he saw the lion's face,
Turned pale, and chilling fear shook all his frame.
He deemed it dangerous to stay or show
His fear by calling, or to try escape.
If he should flee, behold the hungry beast
Long without food would roar and circle him.
If he remained, since this accursed spot
Was filled with devils, he must fear the night.
His strength was gone, his mind distraught by fear,
His limbs forsook him, failed him too his Lord.
The cries from him who shouted in the cave,
Quite grievous and most pitiful, increased.
He promised wealth and lands and fine châteaux,
And everything that heart or lips can pledge.
And now so long he'd begged and promised things
That he had lost his voice and breath alike.
And had a stronger greed than sympathy
Not drawn Bernardus to a kindly act,
Your voice, your life, and all you have, to you,
Dryanus, would have said a last farewell.
But since through greed all dangers are incurred,
The greedy man was made to change his mind.

A parching thirst consumes and hunger grows
More ravishing when hope sees no results.
At length Bernardus flung the ropes, and from
The prison floor drew up his sorry load.
As many different kinds of endings come
To various things, especially the bad,
As he was on the point of stepping out,
And happy to set foot upon the ground,
The rope pulled loose, and back he would have gone
Had not Bernardus quickly rescued him.
He stopped his sudden fall and held him tight,
He pulled him out and to his home restored.
Dryanus thus to land and home and wealth
Returned, now freed from dangers, loss, and death.

Dryanus' Evasion and Perfidy Are Described

SOON AFTER THIS Bernardus came and asked
Dryanus that he do as he agreed.
And noting that his attitude had changed,
He mentioned what he did and pressed his case,
And asked, if he could not agree to all,
That he would give at least some part to him.
But he refused and claimed the man was mad,
And used his dogs to keep him from his house.
And that he might the better make the wretch
Be still, he threatened to cut off his head.
Dumbfounded by these threats Bernardus left
His halls in haste, a finger to his lips.
For knowing well we must have fear of power,
That anger for the rich is often rash,
He thought it better to restrain his tongue
Than to insist Dryanus keep his pledge.
The man suspected when he doesn't speak

Is surely more suspected when he does.
It isn't safe to irritate the rich;
They're seldom loyal and their love is feared.

On the Remuneration Given to Bernardus by the Ape, the Lion, and the Serpent

THE FOURTH DAY went Bernardus to the woods,
And just as usual took Brunellus too,
When lo! the lion with a gentle mien
Brought game, and motioned with his head and paw.
Thus to repay his thanks for kindly deeds,
And pleased to aid in every way he could,
He caught the fattest from a herd of deer
And brought it as a present to his lord.
And when the lion left, the ape arrived
With lowered head, and on her back had wood.
A heap of wood she brought for his reward,
Which she herself had dried, and near him placed.
That she had gathered this for him to take
She graciously revealed by nods and signs.
Not once or twice did they intend to do
These deeds, but on each day Bernardus came.
These two awards of meat and seasoned wood
He got each day he went into the woods.
Nor on the serpent were his efforts lost;
It for the slightest favor gave the most.
One day it humbly came to him as if
Desiring, if it could, to utter thanks.
A gem which lay concealed inside its mouth
It let drop out into his open hand.
And lest by lingering it frighten him
And spoil the value of its gift to him,
It quickly fled, and by its flight gave joy,

And favored more the giver and the gift.
For gifts from him whose sight offends the eye
Give little pleasure, and his words give less.
Such thoughts as these and actions from the beasts
Bernardus, seeing, was amazed and pleased.
This triple gift of lion, ape, and snake
Reveals the noble character of beasts.
When beasts repay good turns with due reward,
They bring reproach on men who show no thanks.
Bernardus proved that he had learned this well,
Since for such aid Dryanus served him ill.

Bernardus Is Made Rich by the Stone Which the Serpent Gives

ENRICHED WITH SUCH A GIFT Bernardus sped
To town, made lighter by a happy heart,
And seeking to find out the jewel's worth
He canvassed all the jewelry shops in town.
The stone was new and never seen before
Inside this city, and its worth unknown.
But he at last, in need of money, took
And traded it for thrice its weight in gold.
Upon returning home it pleased him more
To hold the money than to count it up.
The stone which he just now had sold for gold
He found placed back inside his money bag.
He wished to keep it but dared not, because
The prince's eunuch bought it, whom he knew.
So back he went, pretending he had erred,
And gave it back, made honest by his fear.
Bernardus scarcely to his home returned
When lo! the stone again was in his bag.
He gave it to the eunuch once again,

But on returning home, it too returned.
And now it had returned as many times
As it was taken, faring back and forth.
Report of this came to the ruler's ear:
This story of the stone that thus came back.
And he, disturbed by this strange thing, had brought
Before him both the stone and those concerned.
The summons brought together rich and poor
Who entered and filled up the spacious hall.
At risk of life, of rank, and rights of state,
By faith in Christ and every sacred tie,
The king bade any who might know the truth
To speak and have no fear of punishment,
To tell the hidden reason, why the stone
So often given could not be retained,
And whence it came or in what region found,
What name it had, what glory, if disguised.

Bernardus' Reply to the King on the Acquisition of the Stone

BERNARDUS then alone arose and when
He'd called for silence fell before the king,
And rising up at once, he put aside
His fear, regained his failing strength, and spoke:
"There lately fell into a deep abyss
Four animals, a man and three wild beasts.
Denied to these was all escape, and save
For help from me they never could have left.
Indeed the pit was formed in such a way
That blinding darkness blotted out the day.
While I accompanied by my ass approached
The woods, a dreadful sound fell on my ears.

In seeking what it was and whence it came,
  At last by close attention, I found out.
The voice replied to me, because it was
  Dryanus' who had been there now four days.
Into the pit a lion, ape, and snake
  Had slipped, and he became the trio's guest.
Yet quiet was he, lest if he should divulge
  Their names, he'd frighten me and I would flee.
But as it later happened I, poor wretch,
  Discovered all, as witness my dread plight.
In sore distress and fearful for his life,
  He first pledged riches, then he begged my help.
He swore by God Almighty and his own
  Baptism and the head and feet of Christ,
Saint James and Thomas Cantuariens,
  To whom he even vowed to go barefoot,
That he would share with me his entire wealth,
  If through my efforts I would rescue him.
A love and hope of gain kept urging me;
  Dryanus begged and begged I pity him.
But more, God knows, did pity win my heart
  By prompting me and bringing force to bear.
What I accomplished then by my own wits,
  Along with honest toil, the record tells.
Because the lion, snake, and ape were saved,
  Not I, Bernardus, am to blame, but he.
For since the rest were hid, what plea save his,
  What bribe or reason would have worried me?
The four I rescued, him the last of all,
  Although with them he could have had first place.
Dryanus came out last, yet slow and hard,
  And that not for my slowness, but his weight.
Yes, he was heavy; oh, that he had been

Far heavier, in justice to myself!
If only all my ropes had quickly snapped
And rendered all my skill and efforts vain!
The pit then would have kept the wicked man,
And darkness of the night enveloped his head.
The thoughtful beasts have paid me due respects,
And given larger gifts than were required.
Indeed the lion daily saves me meat
Of game he's killed, and waits until I come.
The ape brings wood and puts it safe aside,
And keeps it for me in a certain place.
I take whatever I desire, and when,
Just like a fixed reward for services.
This stone too from the serpent's mouth I took,
Because of which the present trouble rose.
And if it has some hidden power which will
Not let it stay, God knows I don't know why.
These things are thus, O lord my king, nor do
I think your servant has disguised the truth.
I'd like for you, apart from me, to find
The cause for this and whether it is true.
If I am truthful, let me have rewards;
If false, then let me suffer sword and cross."

The King's Decision for Bernardus against Dryanus

THE KING thus hearing had Dryanus called,
And he came promptly, but denied the deed.
But since the king was certain of the truth,
And since the thing itself had triple proof,
He ordered that Dryanus should agree
That with Bernardus he would share his wealth,
Or spend three days in that well-known pit,
And dwell with those with whom he lived before.

The King's Decision Pleases All

THE KING'S DECISION suited everyone,
And this arrangement pleased Bernardus too.
Dryanus then preferred to share his goods
And also keep his bargain and his pledge
Than yield himself to fortune such as might
By sharing nothing take from him his all.
He therefore sadly and reluctantly
Divided all and gave Bernardus half.
No less for this did he incur the name
Of ingrate and perennial reproach.
Bernardus then was rich; Dryanus wept
And sorrowed deeply since he lost so much.
Behold such treatment ingrates often pay
For not returning due rewards for aid.
It's often better then to offer gifts
With willing heart than hold away the hand.
To pay a due reward for kindnesses
Is most commendable, and God so bids.
Man's work or toil must never go unpaid
Because he shouts not in the ear of God.
A new command says every man deserves
To have full pay according to his work.
So he who gives not as the time and case demand,
Brings very heavy losses on himself.

Here the Author Concludes His Work with an Exhortation To Acquire Understanding

GRANT this may be a warning to all men—
Brunellus' life; the author so intends.
For not the sound of words but what they mean
The reader must examine carefully.
And he can then perhaps decide himself
What pitfalls he may properly avoid.
For there are some who wish to climb the heights
And struggle day and night to reach this goal,
And while they thus aspire and labor hard,
They often quickly fall, and often fail.
To gain what is opposed to nature and to fate
No one is able to accomplish that,
As witness this Brunellus who, though he
Sought foolish things, yet ever stayed the same.
He's lucky then who through another's risks
Is rendered cautious and by reason ruled.
There's more to say, but more I do not wish
To write—these modest verses must suffice.
I humbly beg the reader now, if he
Finds some defect in these, or lack of sense,
He kindly pay no heed and praise me to
The Blessed Son of Mary, whom men pray.

End of the *Speculum stultorum*